Enrico Mugnaini, M.D.
Laboratory of Neuromorphology
Department of Biobehavioral Sciences
Box U-154
The University of Connecticut
Storrs, Connecticut 06268

GROWTH OF THE NERVOUS SYSTEM

GROWTH OF THE NERVOUS SYSTEM

A Ciba Foundation Symposium

Edited by
G. E. W. WOLSTENHOLME
and
MAEVE O'CONNOR

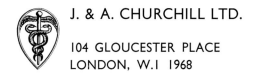

J. & A. CHURCHILL LTD.

104 GLOUCESTER PLACE
LONDON, W.1 1968

First published 1968

With 101 illustrations

Standard Book Number 7000 1353 9

ALL RIGHTS RESERVED

This book is protected under the Berne Convention. It may not be reproduced by any means, in whole or in part, without permission. Application with regard to reproduction should be addressed to the Publishers.

Printed in Great Britain

Contents

Sir John Eccles   Chairman's opening remarks   1

Session 1   Development of specific neuronal connexions

J. Szentágothai	Growth of the nervous system: an introductory survey	3
S. M. Crain E. R. Peterson M. B. Bornstein	Formation of functional interneuronal connexions between explants of various mammalian central nervous tissues during development *in vitro*	13
Discussion	Crain, Eccles, Gaze, Hník, Kerkut, Murray, Sperry, Stefanelli, Székely	31
R. W. Sperry E. Hibbard	Regulative factors in the orderly growth of retino-tectal connexions	41
R. M. Gaze W. E. Watson	Cell division and migration in the brain after optic nerve lesions	53
Discussion	Buller, Drachman, Eccles, Gaze, Hamburger, Hughes, Kollros, Levi-Montalcini, Prestige, Singer, Sperry, Székely, Szentágothai, Walton, Young	67
G. Székely	Development of limb movements: embryological, physiological and model studies	77
Discussion	Drachman, Eccles, Gutmann, Hník, Hughes, Kerkut, Kollros, Piatt, Székely	93
General Discussion	Eccles, Hamburger, Hughes, Sperry, Székely, Szentágothai	96

Session 2   Development of movement

V. Hamburger	The beginnings of coordinated movements in the chick embryo	99
Discussion	Buller, Drachman, Eccles, Hamburger, Hník, Hughes, Kerkut, Kollros, Levi-Montalcini, Mugnaini, Muntz, Székely, Szentágothai, Walton	105
A. Hughes	Development of limb innervation	110
Discussion	Drachman, Eccles, Gaze, Gutmann, Hamburger, Hník, Hughes, Kerkut, Kollros, Levi-Montalcini, Piatt, Prestige, Singer, Székely, Szentágothai, Walton	117

Session 3   Role of chemically specific signals in the development of the nervous system

Rita Levi-Montalcini P. U. Angeletti	Biological aspects of the nerve growth factor	126
Discussion	Crain, Drachman, Eayrs, Eccles, Gutmann, Kerkut, Levi-Montalcini, Murray, Singer, Szentágothai	142

Margaret R. Murray Helena H. Benitez	Action of heavy water [D$_2$O] on growth and development of isolated nervous tissues	148
Discussion	Crain, Gaze, Hník, Kerkut, Levi-Montalcini, Muntz, Murray, Szentágothai	174
J. J. Kollros	Endocrine influences in neural development	179
Discussion	Crain, Eayrs, Hughes, Kerkut, Kollros, Levi-Montalcini, Piatt, Prestige, Stefanelli, Székely, Szentágothai	192

Session 4   Trophic interaction, peripheral and central

M. Singer	Penetration of labelled amino acids into the peripheral nerve fibre from surrounding body fluids	200
Discussion	Drachman, Eccles, Gaze, Hughes, Mugnaini, Murray, Singer, Stefanelli, Szentágothai, Walton, Young	215
G. A. Kerkut	Transport of material along nerves	220
Discussion	Buller, Crain, Drachman, Eccles, Gaze, Gutmann, Hník, Hughes, Kerkut, Mugnaini, Szentágothai, Young	229
E. Gutmann	Development and maintenance of neurotrophic relations between nerve and muscle	233
Discussion	Buller, Drachman, Eccles, Gutmann, Hník, Hughes, Kerkut, Levi-Montalcini, Murray, Prestige, Szentágothai, Walton, Young	243
D. B. Drachman	The role of acetylcholine as a trophic neuromuscular transmitter	251
Discussion	Crain, Drachman, Eccles, Gutmann, Hník, Kerkut, Singer, Székely, Walton, Whittaker	273
General Discussion	Eccles, Gutmann, Hamburger, Hughes, Kollros, Mugnaini, Prestige, Stefanelli, Szentágothai, Walton	279

Author index   289

Subject index   290

Membership

Symposium on Growth of the Nervous System held 21st-23rd June, 1967

Sir John Eccles (Chairman)	Institute for Biomedical Research, Chicago, Illinois
A. J. Buller	Department of Physiology, The Medical School, University of Bristol
S. M. Crain	Department of Physiology, Albert Einstein College of Medicine, Yeshiva University, New York
D. B. Drachman	New England Medical Center Hospitals, Boston, Massachusetts
J. T. Eayrs	Department of Neuroendocrinology, Institute of Psychiatry, The Maudsley Hospital, London
R. M. Gaze	Department of Physiology, University Medical School, Edinburgh
E. Gutmann	Czechoslovak Academy of Sciences, Institute of Physiology, Prague
V. Hamburger	Department of Biology, Washington University, St. Louis, Missouri
P. Hník	Czechoslovak Academy of Sciences, Institute of Physiology, Prague
A. Hughes	Department of Zoology, University of Bristol †
G. A. Kerkut	Department of Physiology and Biochemistry, University of Southampton
J. J. Kollros	Department of Zoology, University of Iowa, Iowa City, Iowa
Rita Levi-Montalcini	Department of Biology, Washington University, St. Louis, Missouri
E. Mugnaini	Anatomisk Institutt, Universitetet i Oslo, Norway
Louise Muntz	Department of Zoology, University of Bristol

† Present address: Department of Anatomy, Western Reserve University School of Medicine, Cleveland, Ohio.

MEMBERSHIP

Margaret R. Murray	Department of Surgery, College of Physicians and Surgeons, Columbia University, New York
J. Piatt	Department of Anatomy, The School of Medicine, University of Pennsylvania, Philadelphia, Pennsylvania
M. C. Prestige	Department of Zoology, University of Bristol.
M. Singer	Department of Anatomy, Western Reserve University School of Medicine, Cleveland, Ohio
R. W. Sperry	Division of Biology, California Institute of Technology, Pasadena, California
A. Stefanelli	Istituto di Anatomia Comparata "Battista Grassi", Università di Roma, Rome
G. Székely	Department of Anatomy, University Medical School, Pécs, Hungary
J. Szentágothai	Department of Anatomy, University Medical School, Budapest, Hungary
J. N. Walton	Regional Neurological Centre, Newcastle General Hospital, Newcastle upon Tyne
V. P. Whittaker	Department of Biochemistry, University of Cambridge
J. Z. Young	Department of Anatomy, University College London

The Ciba Foundation

The Ciba Foundation was opened in 1949 to promote international cooperation in medical and chemical research. It owes its existence to the generosity of CIBA Ltd, Basle, who, recognizing the obstacles to scientific communication created by war, man's natural secretiveness, disciplinary divisions, academic prejudices, separation by distance and by differences in language, decided to set up a philanthropic institution whose aim would be to overcome such barriers. London was chosen as its site for reasons dictated by the special advantages of English charitable trust law (ensuring the independence of its actions), as well as those of language and geography.

The Foundation's house at 41 Portland Place, London, has become well known to workers in many fields of science. Every year the Foundation organizes six to ten three-day symposia and three or four shorter study groups, all of which are published in book form. Many other scientific meetings are held, organized either by the Foundation or by other groups in need of a meeting place. Accommodation is also provided for scientists visiting London, whether or not they are attending a meeting in the house.

The Foundation's many activities are controlled by a small group of distinguished trustees. Within the general framework of biological science, interpreted in its broadest sense, these activities are well summed up by the motto of the Ciba Foundation: *Consocient Gentes*—let the peoples come together.

Preface

Dr. Arthur Hughes first suggested that the Growth of the Nervous System would be an appropriate subject for a Ciba Foundation symposium. It seemed a good time for those who were studying the development of the nervous system and its connexions in the non-human maturing embryo to share their thoughts with scientists working on the transmission of impulses along the nerve fibres to and from specific end-organs in man and other adult animals. Another aim of the meeting was to try to link this basic material with the ways in which the development of the central nervous and neuromuscular systems could go wrong in man. After many pleasant discussions with Arthur Hughes (by letter and in person) and Sir John Eccles (by letter), the meeting took shape.

Sir John was our chairman. This is the second occasion on which he has taken the chair at a Ciba Foundation symposium, and once again his scientific curiosity and his dedication to the nervous system ensured that he performed this function most ably. He was as always a tremendous stimulus to all the members.

We were particularly privileged that Lord Adrian came to some of the sessions. As well as his perpetual interest in and contributions to the physiology of the nervous system he has been (and still is) mentor to several of the members of the symposium.

We would especially like to thank Arthur Hughes for his help in editing these proceedings, and for his non-stop enthusiasm before, during and after the meeting.

We hope that the questions uncovered and unanswered at the symposium and set down in this volume may—perhaps because of their number and complexity—be a challenge to all workers in this field.

CHAIRMAN'S OPENING REMARKS

Sir John Eccles

I am very honoured to be again the Chairman of a Ciba Foundation symposium. The last symposium I chaired here was on the nature of sleep. My problem was of course to prevent participants from giving practical demonstrations! In fact this is the fifth Ciba Foundation conference I have attended, and I can assure those of you who have not been here before that they are most delightful and instructive, friendly and sociable occasions.

I am glad to welcome Lord Adrian here today. He has experienced personally the whole development of investigations on the nervous system at a tremendously exciting period. I want also to thank Dr. Hughes and to say how much I admire the way in which he has developed this conference and helped to choose the various participants.

Now I want to say a few words about the subject of our symposium. You will appreciate that I am quite obsessional about the nervous system and therefore you may discount quite a lot of what I tell you. Maybe you are obsessional too—I hope that you are. First, the fully developed brain, such as mammals possess, is the most complex organization of matter known to occur. The second remarkable thing is that it has just grown. It is not even organized by activity. We know, for example, from Hubel and Wiesel's work that the specific connexions from the retina to the occipital cortex are there before the kitten has opened its eyes and before there has been any retinal stimulation. This extreme complexity has grown without the help of any outside signals and even without any activity of a nervous system playing some role in the neurogenesis. Undoubtedly activity does modulate the later more subtle changes that take place, including the plastic changes in memory, but before that ever starts at all this extremely complex organization of the brain is already in existence.

To study the growth of the brain as such would be an almost insuperable task and most of this conference will be about the simpler aspects of growth and organization such as can be seen, for example, in the peripheral nervous system, or in the simpler animals under controlled conditions. But all of it hangs together in giving us insight until eventually we shall be able to understand how a nervous system is put together or grows. This conference is on the scientific efforts we have made on this seemingly impossible problem. But we can make progress, and a conference like this is a witness to the progress that has been made, not only in the development of hypotheses about how the nervous system grows, but also in respect of the most precise use of new techniques in the testing of these ideas and in the consequent alteration and formulation of new hypotheses and so on. We shall have much discussion in this field in the next three days.

We know intuitively, and also from the pioneer work of Paul Weiss, Roger Sperry and others, that highly specific chemical substances and surfaces must be concerned in guiding growth. There must be in the growing nerve fibre a specific sensing of surfaces and substances so that it smells its way along, finding surfaces that are attractive and others that are repulsive. In speeded-up ciné-photographs of growing nerve fibres one can see most dramatic examples of acceptance and rejection. There are almost emotional reactions by nerve fibres when they find something they love or something that repels them! This is an initial signal system to growth and we shall hear much about it. We must remember that, as in all biological systems, many things can go wrong. Nothing in biology is 100 per cent certain, I suppose. Just as defects occur in crystal growth, so, even in this well-organized chemical sensing of growth and connexions, many things can go wrong, giving the many aberrancies that are seen. Many of these aberrancies are rejected, and one paper here is concerned with this very important problem of the occurrence of defects of organization. The title of this symposium is the *Growth of the Nervous System*, but regressions are also very important. The nervous system does not simply develop linearly in all directions, but many things grow and regress. However, as a result of all of this amazing biological process, we are eventually given a nervous system. The most refined techniques of radioactive tracers and electron microscopy, neurochemistry, tissue transplantation, and immunology itself, provide tools and techniques for us to probe into this mystery of the growth of the nervous system. Behind all this lies the information given by DNA, RNA and so on, which as it were is basic to the construction of these specific surfaces and substances.

That is our target as I see it. I am delighted to be able to give a few minutes of introduction in this way, to get our thinking tuned to this most remarkable enterprise of trying to understand how a nervous system is put together.

GROWTH OF THE NERVOUS SYSTEM: AN INTRODUCTORY SURVEY

J. SZENTÁGOTHAI

Anatomy Department, University Medical School, Budapest

ALTHOUGH several outstanding scientists have attempted to separate development, growth and regeneration by appropriate definitions—which I am not going to discuss, let alone try to improve—these are properties or phenomena of living matter that are linked to each other in countless respects and by considerable overlapping of their essential fields. If this is so with most tissues and organ systems, it is particularly true for the nervous system. This is reflected in the programme of the present symposium. Even a hundred speakers and three weeks instead of three days would be hardly enough to cover the most interesting aspects and questions of these fields. A selection obviously had to be made and I am confident that the formal presentations and particularly the discussions will succeed in carving out from the amorphous bulk of today's information some meaningful shapes and patterns which may also indicate directions for future research.

The object of my introductory survey is to set the stage for the discussions. However, I cannot resist the temptation to exploit the initiative that the organizers have yielded into my hands, and to try to deflect the discussions as far as possible to the problem of how growth and specific nervous functions are interrelated.

The main questions related to development, growth and regeneration of the nervous system were raised systematically and most admirably at a conference held in Chicago in 1949. The introduction to that conference, by Paul Weiss (1950), is particularly worth recapitulating in order to see how clearly the main issues had then been recognized on the "horizontal plane", but also in order to understand how fundamentally our approach has been changed by the new techniques—electron microscopy, unit level physiology, and modern tissue chemistry with its powerful labelling techniques, to mention only a few—which enable us to proceed towards the heart of questions raised much earlier.

Of the six main topics into which the field was divided in 1950, we shall not be much concerned directly here with the first: *differentiation of the cellular material* forming the nervous tissue as a whole and of its units. The mechanisms of (*a*) *allocation* of the tissue materials, (*b*) *transformation* of this material into the nervous primordia, (*c*) differential *proliferation* of the cells in their various regions, (*d*) *migration* of certain groups of cells giving rise to the separation of cellular matrices into specific nuclear masses of neurons, and (*e*) *destruction* and *resorption* of excess cellular material so that the crude bulk of tissue made available by the processes under (*c*) and (*d*) is reduced to the amount required, have all been extensively studied by classical experimental embryology. Parallel with these processes, differences emerge which are at first only latent and potential, but which become apparent as structural or functional—mainly metabolic—differentiations of nerve which are at first only latent and potential, but which soon become apparent as structural or functional—mainly metabolic—differentiations of nerve cells.

This leads directly to the *second* main topic considered by the Chicago meeting: the *development of nerve processes* and the *establishment of contacts*. Here we are today in a much more favourable position, due to considerable refinement of tissue culturing techniques and their combinations with unit level recording and electron microscopy. It is obvious that with these techniques the early latent constitutional differences between various kinds of neuroblasts and the conditions under which they become fully stabilized —whether (and from what time on) this occurs through self-differentiation or depends on inductive influences from the environment—will be unravelled.

The general problem of development of nerve processes can be divided into various subheadings, and among these we have no formal presentation at this meeting on (*a*) the *elementary mechanism* of axon growth. However, axon growth is only a particular—although excessive—case of cell process formation and pseudopodial movement, and up-to-date consideration of these mechanisms would probably have required an entirely different group of participants from the present one. But it might still be relevant to discuss this field—even if only marginally—in connexion with various influences on protein synthesis, nerve growth factors and the transport of materials along nerves.

Whereas at the time of the Chicago meeting the question (*b*) of *orientation* of the outgrowing nerve processes was considered mainly from its general, i.e. non-specific, aspects, such as micellar orientation of the medium in which they grow, guiding surfaces of cells or cellular masses, and so on, our present approach to these questions will certainly be more on the specific side. This is mainly due to the magnificent model of retino-tectal connexions which has now been studied with a large variety of techniques and from different viewpoints. We look forward to the discussion of the factors that give direction to the advance of growing axons, particularly the question of how far these can be explained by non-specific interactions in the above sense, or how far these guiding influences are really specific.

The questions of specificity will arise more sharply when we discuss the mechanisms of (c) *establishment of contacts*. It is obvious from classical neurohistology that some specific selection process governs the establishment of contacts by specific kinds of synapses between certain sets of neurons and the prohibition of contacts between others. Unfortunately, little is known about the chemical or perhaps quasi-immunological mechanism underlying the selective capacity of certain cell membranes—or parts thereof—to form firm associations with certain regions of membranes of other cells. The formation of synaptic contacts seems again to be only a very peculiar case of the general capacity of cells to associate or conversely to dissociate. The importance of this phenomenon in many fundamental biological mechanisms, particularly in cancer research, would make it worth while to focus attention on this aspect of nervous tissue development where it occurs with such spectacular refinement.

The situation, however, becomes much more intricate because inside given sets of neurons there are specific guiding mechanisms that secure the orderly formation of strictly determined projections. Since such determined and refined projection patterns are the rule in most neural connexions, the mechanisms by which they are established are of extreme importance for our understanding of how the complex wiring of the nervous system comes into being. It will therefore be most interesting to discuss these guiding mechanisms, in particular how far they are primarily specific, that is, whether growing axons are specifically conducted to their respective goals, or whether they are brought to their approximate sites of termination only by chance. In the latter case the establishment of orderly connexions, even in very intricate topographical patterns, could be explained by (d) contacts established by a few early *pioneering* fibres, essentially by trial and error, and (e) *fasciculation* by later fibres growing along the pioneering fibres that have undergone some progressive differentiation in consequence of successful contacts (for example by quickly growing in diameter, as shown in the periphery by Aitken, Sharman and Young, 1947).

(*f*) *Saturation factors* and (*g*) *differential timing* of the outgrowth of axons from various topical subgroups of the set of neurons giving rise to the pathway may be additional mechanisms of such guiding systems initially based on retrograde selection. With the cerebellar cortex as a convenient model, where the granule neurons can be destroyed easily by X-irradiation in the newborn animal while the neurons are still in the outer cellular matrix layer, some information on saturation factors has been gained recently. As shown by Dr. Hámori (unpublished), mossy fibres that otherwise are restricted to the granule layer invade the molecular layer after X-irradiation, and even establish there synapses with neurons other than granule cells, when most of the parallel fibres in the molecular layer are lacking. These experiments only repeat the findings of the analogous experiments of Morris (1953) on the sprouting of motor fibres in the partially denervated muscle; however it is significant that this occurs in a central nervous tissue having such a highly specified connectivity and in a relatively late stage of development.

The change between 18 years ago and today is also well characterized by the fact that while the studies on specificity and selectivity and on the development of behaviour (*fifth* and *sixth* main topics at the Chicago meeting) had to be based almost exclusively on behavioural observations, either in normal development or after various experimental interferences, we are now witnessing attempts to correlate the development of various functions with exact histological—to some extent quantitative—and physiological observations. What is more, we are approaching the time when the proposal of simplified neuron circuit models for various functions so that they can be tested in simulation experiments can no longer be rejected as unrealistic science fiction. In spite of the high specificity and strict topographical patterning in which many nervous connexions appear to be established during development, it would be a mistake to base all reasoning about elementary nervous functions on strictly predetermined specific connexions. This would be to revert to a somewhat mechanistic "switch-board" concept of nervous structures. A recent reinvestigation in our laboratory, of spinal cord structure as revealed by Golgi staining methods supplemented by combined electron microscope and degeneration studies, reduces confidence in an "all-too-specific" neuron circuit model and leads us to favour the concept of a central core neuron network with a considerable degree of randomness in its connectivity. The immense number of initial collaterals given by most interneurons and the distribution of their endings over wide areas of the grey matter show how small a fraction of synaptic terminals per neuron may belong to the pathways which have rather strictly determined, almost point-to-point connexions. It is readily admitted that, considered in statistical terms, this apparent "randomness" on a larger scale may be part of a very specific pattern. Nevertheless, it is important to investigate the properties of small groups of mutually interconnected neurons (elementary neuron networks) because they have functional capacities very different from those of neurons linked simply in divergent or convergent chains. Even the simplest networks of neurons with mutual and recurrent interconnexions that have so far been investigated by simulation with models (Harmon, 1964; Székely, 1965) exhibit an unexpected functional versatility. As the neurons in many parts of the central nervous system—particularly the spinal cord—are in fact connected in this way, it would be a mistake to neglect this important aspect and the approach by models.

Undoubtedly the central topic of the Chicago conference was the *third*: the *growth proper* of nervous elements. Here one has to keep in mind that while the early differentiation of nerve tissue, the development of nerve processes and even the establishment of synaptic contacts are largely "prefunctional", the phenomena of growth and the complex trophic interactions linked with them are "subfunctional". There is much evidence to show that specific functions—impulse generation, conduction and transmission—have a very significant repercussion on structural-functional parameters of nerve elements that have to be classified in the larger framework of growth or, in the negative direction, of regression. These phenomena were most thoroughly and systematically investigated by J. Z. Young and co-workers over a number of years immediately after World War II. The most relevant conclusion reached (Young, 1948) was the concept of "double dependence" of growth processes, on the stimulation received and on that transmitted forwards. Using other models, we were later able to show (Szentágothai and Rajkovits, 1955) that (*a*) the same kinds of changes can be induced by deprivation of function as are caused by interruption of the neuron chain at a distant link, and (*b*) that such changes affect various parts of the neuron at once and in the same direction, namely withering of the dendritic tree, reduction of cell size, of nucleolar and nuclear size, reduction of the axon diameter and myelin sheath thickness, and finally the shrinking of synaptic terminals. It is obvious that these various changes are manifestations of the same intrinsic change in the neurons, probably in connexion with changes in their mechanisms of protein synthesis. Although for practical reasons most of these changes have been investigated only in the negative direction—that is, using functional deprivation—there is good evidence (Edds, 1950) that the mechanisms work equally well and with opposite sign when there is functional overloading. A more recent study by Wiesel and Hubel (1963*a*, *b*) on correlations between structural and functional changes in the optical system during visual deprivation of the growing animal not only substantiates earlier assumptions based on crude histological observations but also indicates that functional connexions that existed at birth become disrupted. Thus, when we discuss transport of substances through the sheath of the axon or along the nerves and particularly the maintenance of neurotrophic relations and the role of hormones and neurotransmitters in neurotrophic mechanisms, it would seem advisable to keep always in mind the possible repercussion of specific function upon these mechanisms.

This is all the more important as many authors have felt (Ramón y Cajal, 1911; Hebb, 1949; Eccles, 1953, 1966; Young, 1951, 1964) that the mechanisms of memory might be related to processes of neuron growth, especially at the level of the synapse. It was to be expected that after the first crude attempt to detect changes in size of synaptic clubs (Szentágothai and Rajkovits, 1955) under the light microscope, interest in this respect would soon turn towards dendritic spines, the synaptic relations of which can be studied so well in the electron microscope (Gray, 1959, 1961). It was not long before secondary branches of spines were observed (in the cerebellar cortex: Hámori and Szentágothai, 1964) which established synapses with an additional terminal axon passing close to the main spine. As such secondary spines were found in abundance in some cases but not in others, the question was tentatively raised of whether this might be a growth mechanism by which the number of contacts and consequently the effectiveness of the synapse could be increased. Still more important are observations of so-called microspines or spinules in various regions of the central nervous system, protruding from post-synaptic surfaces of dendrites or cell bodies and intruding into invaginations of the presynaptic axonal surfaces (M. Colonnier, personal communication; Hámori and Szentágothai, unpublished). These spinules are also abundant in certain synapses of

---

FIG. 1. Structure of glomerular synapse in the dorsal lamella of the lateral geniculate body in a two-month-old dog.

The optic axon terminals (Opt) can be recognized and distinguished from numerous non-optic (No) axon terminals on the basis of their large empty mitochondria (M). Both kinds of axons establish synapses with club-shaped protrusions of dendrites (Dp) entering the synaptic glomeruli. Spines (Sp) of these dendritic protrusions are embedded in deep invaginations of the optic axon terminals and they establish synaptic (Sy) contact zones of the usual structure. No spines ever invade or contact the non-optic axons. The glomerulus is surrounded by a glial envelope (Gl). The lower photograph shows three invaginated spine profiles in a single optic axon terminal, two of which have synaptic contact zones.

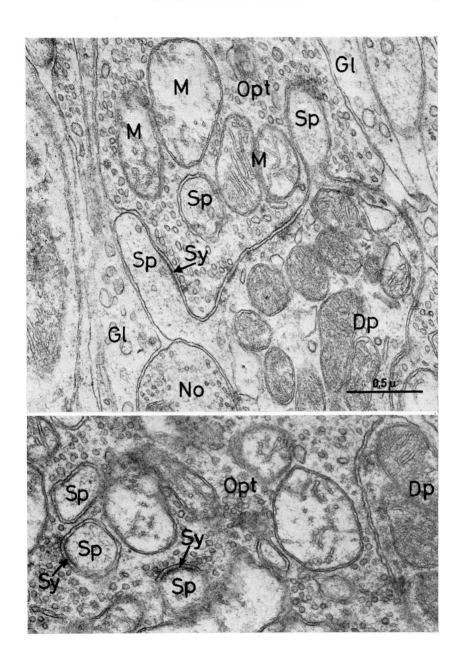

one animal but rare or lacking in the same synapse of another. An obvious explanation would be to assume that they are not fixed structures, but vary with functional circumstances. Most recently Valverde (1967) has provided convincing evidence that in certain dendritic regions of the primary visual cortex the number of spines is reduced in animals reared in complete darkness. We have made a comparable observation recently in the lateral geniculate body of the dog. Here we are taking advantage of (*a*) the possibility of identifying the synaptic terminal of optic fibres among many other axon terminals on the basis of their large empty mitochondria (Szentágothai, Hámori and Tömböl, 1966), and (*b*) the spines that protrude in considerable numbers from the dendritic ends (so-called dendrite protrusions) into the optic endings (Fig. 1), but not into others. These are true spines and not spinules as they have real synaptic contacts with the optic endings. If the eyelids of the dog are sutured soon after birth (which is not the same as rearing an animal in complete darkness, since some light stimulation through the eyelids is unavoidable), these dendritic spines do not develop or are extremely rare (Fig. 2). Thus deprivation of natural stimuli during growth of the animal may indeed prevent appropriate structural elaboration of synapses.

One should be cautious, however, in exploiting such an observation too much in favour of the growth hypothesis of memory, particularly against a protein hypothesis (Hydén, 1964). First of all, the growth hypothesis would obviously also require changes in protein synthesis. It is remarkable that specific post-synaptic regions of dendrites—the so-called dendritic protrusions or digits from which the spines may grow out—quite often contain ribosomes or polysomes. But, although I firmly believe that such phenomena of growth must be very closely linked to the "more peripheral" part, if one may say so, of memory—that is, the points where the elementary traces of stimulation events become "incorporated" or, conversely, where they are "read out"—I would have great difficulty in accepting the notion that the traces are in reality fixed at these sites. The human brain can recall events of 50 years or more ago, even under circumstances in which there is no external reinforcement of the original stimulus pattern. To suppose that changes in the number or size of spines and various other kinds of synaptic contacts could be responsible for this is well beyond the limits of imagination. So far we know that the chromosomal nucleic acids have the unique capacity of retaining encoded information indefinitely. The most promising way of proceeding would therefore be to try to bridge the gap between the synapse and chromosomal nucleic acids. This gap is, of course, not bridged by frequency concepts (Hydén, 1964)

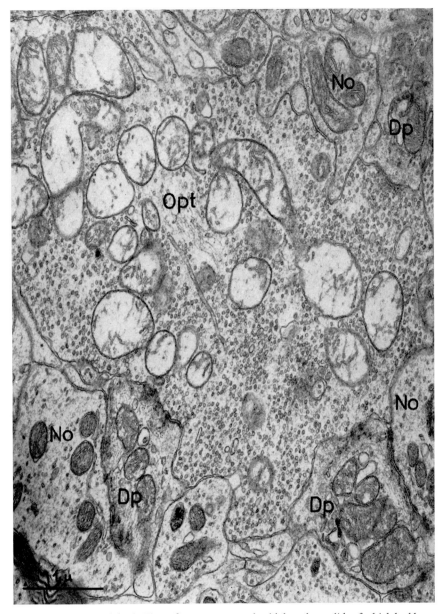

Fig. 2. Same material as in Fig. 1, from a two-month-old dog, the eyelids of which had been sutured on the 7th day after birth.

Large optic axon terminal (Opt), although in synaptic contact with several dendritic protrusions (Dp), shows no invaginated spines. Intraglomerular spines are lacking in this material. Non-optic (No) axon terminals exhibit the features—fewer, smaller and ovoid synaptic vesicles—that are considered by many authors as characteristic of inhibitory synapses.

because these cannot be reconciled with the basic facts of neurophysiology (Eccles, 1966).

SUMMARY

A brief survey is given of some of the main topics in the related fields of development and growth of the nervous system and of the changes in approach and outlook that their study has undergone in the last two decades. While developmental processes are largely prefunctional, those of growth are inseparably linked with specific nervous functions. Growth or quasi-growth processes induced or influenced by nervous functions are briefly discussed and their importance is emphasized in view of a possible relation between minute adjustments of neuron growth and the basic mechanisms of memory.

REFERENCES

AITKEN, J. I., SHARMAN, M., and YOUNG, J. Z. (1947). *J. Anat.*, **81**, 1–22.
ECCLES, J. C. (1953). *The Neurophysiological Basis of Mind.* Oxford: Clarendon Press.
ECCLES, J. C. (1966). In *Brain and Conscious Experience*, pp. 314–344, ed. Eccles, J. C. New York: Springer-Verlag.
EDDS, M. V. (1950). *J. comp. Neurol.*, **93**, 259–275.
GRAY, E. G. (1959). *J. Anat.*, **93**, 420–433.
GRAY, E. G. (1961). *J. Anat.*, **95**, 345–356.
HÁMORI, J., and SZENTÁGOTHAI, J. (1964). *Acta biol. hung.*, **15**, 95–117.
HARMON, L. D. (1964). In *Problems in Neural Modeling*, pp. 2–30, ed. Reiss, R. F. Stanford: Stanford University Press.
HEBB, D. O. (1949). *The Organization of Behaviour.* New York: Wiley.
HYDÉN, H. (1964). *Bull. Neurosci. Res. Program*, **2**, May–June, pp. 23–38. Brookline, Mass.
MORRIS, D. D. B. (1953). *J. Bone Jt Surg.*, **35B**, 650–660.
RAMÓN Y CAJAL, S. (1911). *Histologie du Système Nerveux*, II. Paris: Maloine.
SZÉKELY, G. (1965). *Acta physiol. hung.*, **27**, 285–289.
SZENTÁGOTHAI, J., HÁMORI, J., and TÖMBÖL, T. (1966). *Expl Brain Res.*, **2**, 283–301.
SZENTÁGOTHAI, J., and RAJKOVITS, K. (1955). *Acta morph. hung.*, **5**, 253–274.
VALVERDE, F. (1967). *Expl Brain Res.*, **3**, 337–352.
WEISS, P. (1950). In *Genetic Neurology*, pp. 1–39, ed. Weiss, P. Chicago: University of Chicago Press.
WIESEL, T. N., and HUBEL, D. H. (1963*a*). *J. Neurophysiol.*, **26**, 978–993.
WIESEL, T. N., and HUBEL, D. H. (1963*b*). *J. Neurophysiol.*, **26**, 1003–1017.
YOUNG, J. Z. (1948). *Symp. Soc. exp. Biol.*, **2**, 57–74.
YOUNG, J. Z. (1951). *Proc. R. Soc. B*, **139**, 18–37.
YOUNG, J. Z. (1964). *A Model of the Brain.* Oxford: Clarendon Press.

FORMATION OF FUNCTIONAL INTERNEURONAL CONNEXIONS BETWEEN EXPLANTS OF VARIOUS MAMMALIAN CENTRAL NERVOUS TISSUES DURING DEVELOPMENT *IN VITRO*†

S. M. CRAIN‡, E. R. PETERSON and M. B. BORNSTEIN‡

Departments of Physiology and Neurology, Albert Einstein College of Medicine, Yeshiva University, New York

EXPLANTS of embryonic mammalian central nervous tissues can develop and maintain organotypic bioelectric activities during maturation in culture (Crain, 1966). Development of these complex functions is correlated with the formation of abundant synaptic junctions between neurons in explants of foetal rat spinal cord (Crain and Peterson, 1964, 1965; Bunge, Bunge and Peterson, 1965) or of neonatal mouse cerebral cortex (Crain and Bornstein, 1964; Pappas, 1966). When the explant includes different types of tissues which would normally develop synaptic relations at later stages *in situ*, such connexions may also form and show characteristic functions after explantation in culture, e.g. spinal cord with attached dorsal root ganglion (Crain and Peterson, 1964; Peterson, Crain and Murray, 1965) or skeletal muscle (Crain, 1964a, 1966; Corner and Crain, 1965); cerebral cortex with attached subcortical tissue (Crain, 1965).

Close correlation of electrophysiological and electron microscope analyses of foetal rat spinal cord cultures has provided strong evidence that synaptic networks can develop even after explantation of immature "synapse-free" tissues (Crain and Peterson, 1967; Bunge, Bunge and Peterson, 1967). In the 14-day foetal rat cord explant, complex bioelectric activities and characteristic synaptic junctions appear for the first time at about three days *in vitro* and show further signs of maturation during the following week in culture. However, since the general topography of the

---

† Supported by research grants NB-06545 and NB-06735 from the National Institute for Neurological Diseases and Blindness, U.S. Public Health Service, and MS-433 from the National Multiple Sclerosis Society.

‡ Kennedy Scholar at the Rose F. Kennedy Center for Research in Mental Retardation and Human Development (Albert Einstein College of Medicine).

neurons and glial cells within this tissue is not radically disturbed by the explantation procedure, the patterns of cellular interrelationships may already be sufficiently developed for synaptogenesis after explantation to be automatically determined.

It was still a moot question, therefore, whether neurons could form functional synaptic relations in culture even under conditions where the neurons had been completely separated at explantation. Axo-dendritic synapses had already been detected, by Bunge, Bunge and Peterson (1965), within the neuropil which formed *de novo* over the exposed, cut surface of relatively mature rat spinal cord explants, but it was difficult to demonstrate selectively the functional capacity of these junctions, in view of their close proximity to the synaptic network within the original explant. Moreover, these newly formed synapses developed between collateral branches of neurons which may have been already interrelated within the main body of the explant. In the present study we have approached this question by culturing pairs or groups of CNS explants separated by gaps of the order of 1 mm. It will be shown that neurites from one explant, after growing across this gap, can indeed make functional interneuronal connexions with neurons within the other explant. These experiments demonstrate the feasibility of using ordered arrays of CNS explants as a model system to supplement and extend studies made *in situ* of interactions among various types of CNS neurons, especially during growth and regeneration.

METHODS

The techniques used to prepare the foetal CNS tissue cultures have been reported in detail (Peterson, Crain and Murray, 1965; Bornstein, 1964). Pairs (or groups) of explants (about 1 mm.^{3}) of foetal rat or mouse spinal cord-ganglia, brain stem, and cerebral tissues were carefully positioned on the surface of collagen-coated cover-glasses (and were oriented so that apposing surfaces were free of meninges). Gaps between explants used in this study ranged up to 1·5 mm. (Figs. 1, 3, 5X, 6X, 7X). The cultures were maintained at 35°C, in Maximow depression-slide chambers, as "lying-drop" preparations.

Electrophysiological methods applied to these explants have been previously described (Crain, 1956; Crain and Peterson, 1964). Cultures selected after serial microscopic examinations, during days or weeks of incubation in Maximow slides, were transferred to a special heated, moist chamber for bioelectric studies (lasting several hours). Openings in the wall of this chamber permitted micromanipulation of microelectrodes for focal recording and stimulation within an explant, under direct visual control, at high magnification (see also Fig. 2).

RESULTS

Electrophysiological evidence of the formation of functional interneuronal connexions between separate explants of various foetal rodent CNS tissues have been obtained in more than 20 cultures after maintenance for one to six weeks *in vitro*. In dozens of other cultures of paired CNS explants, no clear-cut coupling could be demonstrated although complex bioelectric activity could be evoked within each explant.

Spinal cord–brain stem connexions

Preliminary experiments indicated that pairs of foetal rat spinal cord explants could become functionally coupled after the formation of neuritic bridges across gaps of 1 mm., during the first week *in vitro* (Crain and Peterson, 1965). Attempts were then made to determine whether tissues explanted from *different* CNS regions would also develop these interrelationships in culture. Pairing of foetal rat spinal cord and brain-stem explants showed that neuritic bridges could form similar to those in coupled cord cultures (Crain, 1966; Crain and Peterson, 1966). The neurites and glial cells often spread diffusely between the explants, requiring careful study under high magnification to detect the existence of neurites which had bridged the gap (Fig. 1). In some cases, on the other hand, a densely packed neuritic–glial bridge formed between the explants (Fig. 3). The orderliness of the latter type of connexion was often accentuated by slight retraction of the explants away from each other *after* the neurites had spanned the gap.

Simultaneous microelectrode recordings from these paired spinal cord–brain-stem explants demonstrate that spontaneous discharges arising in one explant may be *regularly* followed, after a latency ranging from several msec. to several hundred msec., by long-lasting discharges in the other explant (Figs. 2A,D,G; $4A_{1,2}$, B; and $5B_{4,5}$). The synchrony between the onset of activity in each explant does not generally apply to the temporal pattern or duration of the discharges in the two tissues. Selective stimulation of dorsal root ganglia (DRG) attached to these cord explants (under careful visual control; Fig. 1) permits clear-cut (synaptic) excitation of cord neurons without direct application of a stimulating electrode to the vicinity of the cord explant (Crain and Peterson, 1964). This procedure precludes fortuitous antidromic excitation of brain-stem neurites which had bridged the gap and may then have arborized within the cord explant. Under these conditions, a brief DRG stimulus evokes an early discharge in the cord explant, and this may be followed by a longer-lasting discharge (about 1 sec.) in the brain-stem explant (Figs. 2B, $4A_3$, and 5D). The latency between the onset of activity in the two explants may vary widely, as occurs during spontaneous discharges, and is probably due to gradual activation of brain-stem neurons through polysynaptic pathways from cord neurons (or *vice*

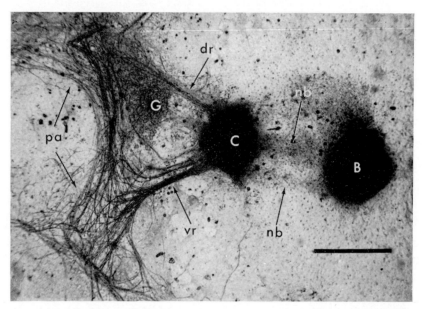

FIG. 1. Living, unstained culture containing an explant of rat spinal cord (C) with attached dorsal root-ganglion (G) located about 1 mm. from an explant of rat brain stem (B).

The tissues were obtained from a 16-day foetal rat and maintained for six weeks *in vitro*. Note long dorsal (dr) and ventral (vr) roots connected to the cord and the peripheral arborizations (pa) of these neurites. Also note dense bands composed of neurites (and supporting cells) which have formed a "bridge" (nb) across the gap between the cord and brain-stem explants. (Neurites in "bridge" could be clearly seen at higher magnification.) Scale: 1 mm.

versa when the brain-stem explant triggers cord: e.g. Figs. 2A, E, and 4B). Spontaneous discharges appeared to occur more commonly in brain-stem than cord explants, in the normal culture medium (Figs. 2A and 4B), although the cord tended to become the pacemaker during strychnine administration (Figs. $4A_{1, 2}$; $5B_5$; cf. Fig. $5B_4$). Moreover, the functional connexions between cord and brain-stem explants were often not detectable until strychnine was introduced into the culture medium, even though complex activities could be generated within each explant. This suggests

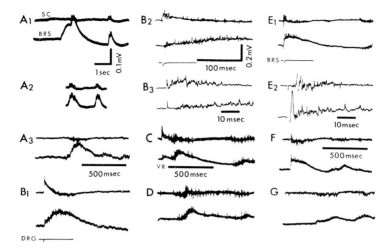

Fig. 2. Functional interneuronal connexions between separate explants of foetal rat brain stem and spinal cord-ganglia after maturation in culture.

Simultaneous recordings from spinal cord (SC; upper sweep) and brain-stem (BRS) explants, connected only by *de novo* neuritic bridge across 1 mm. gap (Fig. 1: nb). ($A_{1,3}$): Spontaneous slow-wave potentials occurring synchronously in both explants. Activity appears to arise earlier in the brain-stem explant (A_3). ($B_{1,2}$): Long-lasting spike-barrage and slow-wave potentials evoked in both explants by stimulus to dorsal root ganglion (DRG) attached to spinal cord explant (located 1 mm. away from the cord, in direction opposite to brain stem). Note similarity between response evoked in brain stem via DRG-cord pathway and the spontaneous discharge (B_1 and A_3, lower sweeps). Also note 5-msec. latency in onset of brain-stem barrage as compared to early appearance of cord activity (B_3). (C): Stimulus to ventral root (VR; 500 μ from edge of cord) evokes similar (but smaller) cord response as with DRG stimulation, and latency of major brain-stem discharge is now greater than 100 msec. Note resurgence of activity in both explants after 400 msec. silent period (C) and spontaneous appearance, shortly afterwards, of discharges (D) similar to those evoked by cord stimulation. ($E_{1,2}$): Stimulus to small group of brain-stem neurites (in growth-zone, about 50 μ from edge of explant) also evokes long barrages in both explants, but now the *cord* response shows a minimum latency of several msec. (compare E_2 to B_3). (F): Similar cord and brain-stem responses are elicited by stimulus applied within brain-stem explant (remote from neuritic "bridge"). (G): Spontaneous discharges again showing synchrony between the two explants.

Note. In this and all subsequent figures: time and amplitude calibrations, and specification of recording and stimulating sites, apply to all succeeding records, until otherwise noted; upward deflection indicates negativity at active recording electrode; all recordings made with silver-core pipettes (25 μ in diameter) and all stimuli applied via saline-filled pipettes (10 μ in diameter), unless otherwise noted; electric stimuli were 0·1–0·3 msec. in duration, and onset is indicated, where necessary, by sharp break in baseline below each pair of sweeps.

that inhibitory pathways may be involved in the coupling between explants (see Figs. 7 and 9), although inhibitory circuits within each explant could probably account for these effects of strychnine.

The essential role of neuritic pathways in mediating the interactions between these separated CNS explants is demonstrated by the complete abolition of transmission from one explant to another after microsurgical

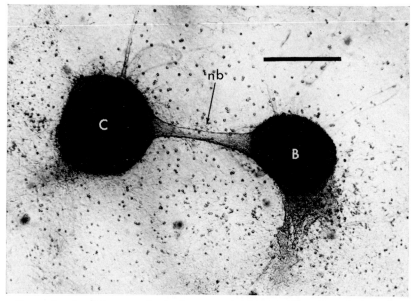

FIG. 3. Another living, unstained culture containing an explant of rat spinal cord (C) located about 1 mm. from an explant of rat brain stem (B).

The tissues were obtained from a 16-day foetal rat and maintained for about 1 month *in vitro*. Note more prominent neuritic bridge (nb) which has formed between the explants, in contrast to diffuse bridge in Fig. 1. Dorsal root ganglion is out of field, located about 1 mm. away from cord, in direction opposite to medulla. (Bioelectric properties were similar to those shown in Figs. 2 and 4.) Scale: 1 mm.

transection of the neuritic bridge across the gap (Figs. 3 and 4C). Additional controls have been made in a number of cultures to clarify the possible role of non-neural spread of current. In the first place, these bioelectric interactions between the explants are not seen in cases where neurites have not grown across the gap—even when the tissues are located less than 100 μ apart (although widespread complex activity may be evoked within each explant). Moreover, after acute midline transection of an explant, electric stimuli applied to the neural tissue on one side of the slit produced the usual bioelectric responses within the confines of the surgical boundary, but no activity could be detected in the tissue beyond the incision (even when the gap was less than 10 μ wide). Controls demonstrating the absence of significant spread of applied stimulating currents have also been made systematically (Figs. 5E$_3$ and 6C).

Cerebral connexions with medulla or spinal cord

On the basis of the favourable results in producing *de novo* coupling of spinal cord and brain-stem explants in culture, various more complex

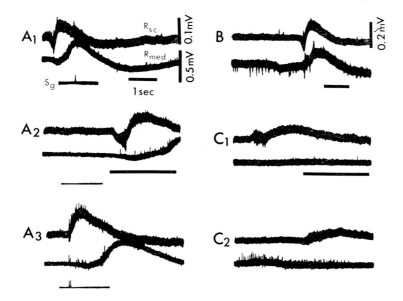

FIG. 4. Effect of surgical transection of the neuritic bridge between functionally coupled CNS explants (similar to array in Fig. 3).

($A_{1, 2}$): Simultaneous recordings of spontaneous activity of spinal cord (R_{sc}: upper sweep) and medulla (R_{med}) explants connected by neuritic bridge across 300 μ gap (after application of strychnine at 10 μg./ml.; R_{med} obtained with 7 μ diameter saline-filled pipette). The tissues were obtained from a 14-day mouse foetus and cultured for 12 days *in vitro*. A stimulus (S_g) was applied *after* onset of spontaneous discharges and was ineffective (cf. A_3). Note synchronization of the slow waves generated by these explants and the long, but regular, latency of the medulla wave after the onset of cord activity. (A_3): Application of stimulus to dorsal root ganglion, located about 400 μ from edge of cord (in direction opposite to medulla), evoked similar cord-medulla discharge sequence as observed spontaneously (but initial negative phase of spontaneous cord slow wave does not occur). (B): Synchronized spontaneous activity is still present after strychnine has been replaced by control medium. Note that *medulla* discharge now arises first (cf. $A_{1, 2}$). (C_1): Shortly after microsurgical transection of neuritic bridge between the cord and medulla explants. Spontaneous (as well as electrically evoked) discharge in cord no longer triggers medulla, and discharge in latter can occur independently of cord activity (C_2).

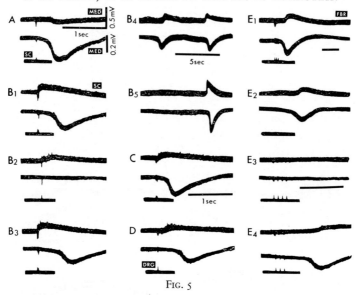

FIG. 5

FIG. 5. Functional connexions between separate explants of foetal mouse forebrain (FBR), medulla (MED), and spinal cord-ganglia (SC-DRG) after maturation in culture. (X): Actual array of explants (from 14-day foetus; three weeks *in vitro*). Microscopic study at high magnification revealed that neurites had bridged gap between cord and medulla, and between medulla and forebrain, but no signs of direct connexions between cord and forebrain explants were detected. Scale: 1 mm.

(A): Simultaneous recordings of responses evoked at two sites in medulla explant (under strychnine at 10 μg./ml.) by single stimulus applied to spinal cord explant (about 1·5 mm. away—see X: "a"). (Records in upper sweeps obtained with 5 μ diameter saline-filled pipette.) Note long latency of medulla responses. (B_1): After transfer of 5 μ diameter electrode from medulla to cord explant (200 μ from site of cord stimuli), response evoked in cord can be seen to occur with very short latency compared to that in medulla. Note that the potential evoked in cord is negative in polarity whereas the medulla response, at this site, is positive (cf. Figs. 2 and 4). (B_2): Single cord stimulus elicits smaller cord response, after longer latency (as compared to response evoked with paired stimuli, at about 20 msec. test interval, in B_1), and no activity develops in medulla explant. (B_3): Application of paired cord stimulus again leads to medulla discharge, but latency is now even longer (almost 0·5 sec. to onset of major positive wave). ($B_{4, 5}$): Spontaneous slow waves occurring synchronously in medulla and cord explants—similar to those evoked by cord stimuli, but showing that medulla discharge can trigger the cord when activity arises first in medulla (B_4: early wave). (C): Similar cord and medulla discharges as in B evoked after transfer of stimulating electrode to far end of cord explant (see X: "b"). (D): Similar cord and medulla responses as in B and C even after selective stimulation of an attached dorsal root ganglion, about 500 μ from edge of cord explant (see X: "DRG"; another ganglion is located under the label "a"). (E_1): After transfer of 5 μ diameter electrode from cord to forebrain explant, DRG stimulation evokes similar medulla response as in D, and this activity is followed, after a still longer latency, by a negative slow wave in the forebrain explant. (E_2): Spontaneous slow waves occurring synchronously in forebrain and medulla, similar to those evoked by DRG stimuli. Note that medulla discharge leads forebrain by more than 100 msec. (E_3): Complete absence of DRG-evoked responses in forebrain and medulla after slight withdrawal of stimulating electrode from direct contact with dorsal root ganglion. (E_4): Restoration of forebrain and medulla responses to DRG stimuli after return of stimulating electrode to close contact with DRG. Latency of forebrain response following that of medulla is seen more clearly at this faster sweep rate.

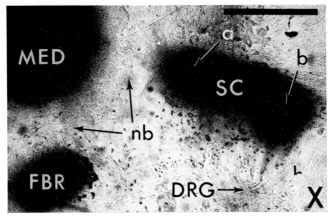

Fig. 5 (X)

arrays were studied, with special emphasis on connexions to cerebral explants. Preliminary attempts have been made to orient various CNS explants so as to facilitate the detection of morphological or bioelectric signs of specificity in the formation of connexions between neurons from one type of explant to another. The data have been ambiguous as far as specificity is concerned, primarily because of wide variability in the patterns of neuritic outgrowth from these CNS explants under our present conditions of culture. Variations with respect to different regions of an explant, in the degree of surgical trauma associated with isolation in culture, and in the rate of neuritic attachment to the collagen-gel substrate could seriously impede detection of any preferential growth of neurites that may occur between specific types of CNS neurons.

Nevertheless, several interesting types of functional interactions have been observed during electrophysiological studies of couplings to cerebral explants. As an extension of the cord–brain-stem experiments, explants of foetal mouse forebrain (excluding cerebral cortical tissue) have been added to the previous array (Fig. 5X). A DRG stimulus now evokes a series of sequential discharges, first in the attached spinal cord, then in the *de novo* coupled medulla (Fig. 5D) and, finally, in the similarly coupled forebrain (Fig. $5E_{1, 4}$). Synchronized spontaneous discharges also occur between all three explants (Figs. $5B_{4, 5}$; E_2), with cord or medulla acting as pacemakers.

Explants of neonatal mouse cerebral cortex frequently generate widespread, synchronized oscillatory (about 5 to 15/sec.) after-discharges, lasting about 1 sec., when triggered by a single brief stimulus (Crain, 1964b, 1966). Local stimulation of a cerebral explant which is coupled to a spinal cord explant (Fig. 6X) evokes a characteristic long-lasting repetitive dis-

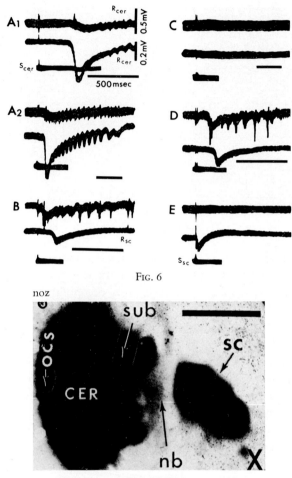

Fig. 6

Fig. 6 (X)

Fig. 6. Cerebrospinal connexions between separate explants of foetal mouse cerebrum (CER) and spinal cord (SC) after maturation in culture. (X): Actual array of explants (from 15-day foetus; 2 weeks *in vitro*). Note that the explants were arranged to facilitate development of neuritic bridge (nb) between cord and subcortical tissue (sub); original cortical (pial) surface (OCS) is at opposite end of cerebral explant. Scale: 1 mm.

(A_1): Responses evoked in "superficial" (upper sweep) and "deep" cerebral cortical tissue by stimulus applied to another deep cerebral site (after application of strychnine at 10 μg./ml.). (Records in upper sweeps obtained with 5 μ diameter saline-filled pipette.) (A_2): Similar responses recorded at slower sweep rate, showing long-lasting complex oscillatory after-discharges at these cerebral sites. (B): Application of stimulus near superficial cerebral recording site (OCS) evokes similar discharge as in A_2 (upper sweep) and simultaneous record in spinal cord explant (about 2·5 mm. away) shows onset of positive slow wave after long latency. (C): Complete absence of evoked responses in cerebrum and cord explants after slight withdrawal of stimulating electrode from direct contact with cerebral tissue. (D): Restoration of cerebral and cord responses even after relocation of stimulating electrode to neurites in outgrowth zone, about 300 μ from OCS edge of cerebral charge in the cerebral tissue (Fig. 6A $_{1,\,2}$), whereas the much simpler slow-wave cord response arises during the course of this cerebral activity after a long latency (Fig. 6B). Selective stimulation of neurites in the outgrowth zone at the "pial" edge of the cerebral explant (Fig. 6X: noz-⊛) indicates that functional corticospinal pathways have probably formed in culture (Fig. 6D), although the neurites which actually bridged the gap between the explants may have been subcortical in origin. Spinocerebral connections had apparently not developed in this case (Fig. 6E) but, in other cultures, locally evoked and even spontaneous cord activities have triggered long-lasting cerebral oscillatory after-discharges.

More complex bioelectric interactions have developed after coupling between explants of foetal mouse cerebrum and medulla (Fig. 7X). The medial portion of the rostral medulla was used for these experiments, in an attempt to develop a model for studies of the effects of the reticular activating system on cerebral tissues. Not only can medulla discharges trigger characteristic cerebral evoked potentials (Fig. 7B$_3$), but repetitive stimulation of the medulla explant appears to produce, at times, a *sustained* increase in the excitability of the cerebral explant (Fig. 7C$_1$). Similar changes in the cerebral tissue, lasting for at least several minutes after repetitive stimulation of a coupled medulla or spinal cord explant, have also been observed without the application of strychnine (see also Crain, 1966).

Facilitation effects are readily demonstrable in these cerebral–medulla explants. Local stimuli which produce long-lasting medulla discharges—but only slight responses in the cerebral explant (Fig. 8A$_{1,\,3}$)—can be temporally summated to trigger large cerebral evoked potentials after long latencies (Fig. 8A$_{2,\,4}$). Furthermore, *spatial* facilitation can be elicited by pairing subthreshold medulla and cerebral stimuli even when the latter follow the former by intervals as long as 75 msec. (Fig. 8B).

In addition to the evidence described above for excitatory interactions between coupled CNS explants, the patterning of spike barrages recorded from the medulla concomitantly with cerebral after-discharges suggests that *inhibitory* impulses may also be transmitted from one explant to the other. Although the data supporting this interpretation are still fragmentary, they indicate that a remarkable degree of complex interaction can occur between these two types of CNS explants. The "silent periods" in the

---

explant (see X: noz-⊛) Response latencies are now greater than in B. (E): Failure of cord-evoked activity to excite cerebral explant. Note that although cord response now occurs after much shorter latency, the amplitude of the slow-wave component is not significantly larger than that evoked by a distant, superficial cortical stimulus (cf. B,D).

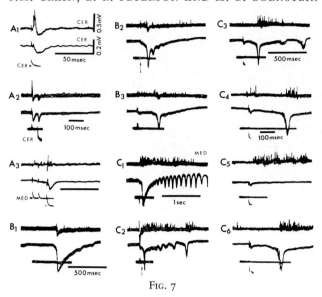

FIG. 7

FIG. 7. Functional connexions between separate explants of foetal mouse cerebrum (CER) and medulla (MED) after maturation in culture. (X): Actual array of explants (from 14-day foetus; 3 weeks *in vitro*). Note that neuritic bridge (nb) has formed between medulla and subcortical edge (sub) of cerebral explant (as in Fig. 6X). Scale: 1 mm.

(A_1): Evoked potentials of relatively short duration, at two superficial cerebral sites, after deep cerebral stimulus (records in upper sweeps obtained with 7 μ diameter saline-filled pipette.) (A_2): Application of paired stimuli at two deep cerebral sites indicates no evidence of facilitation effects. (A_3): Spike potential of brief duration and small amplitude is evoked by stimulus to medulla explant, and little sign of facilitation occurs when test stimuli are then applied at medulla and cerebral sites. (Note that stimulus signals to medulla and cerebral explants appear on 3rd and 4th sweeps, respectively.) (B): Onset of more complex discharges, of larger amplitude and longer duration, after application of strychnine (10 μg./ml.), occurring spontaneously (B_1) and after a cerebral stimulus (B_2). (B_3): Complex cerebral potentials evoked, for first time, by *medulla* stimuli (cf. A_3), after relatively long latency (cf. B_2). (C_1): Much longer-lasting characteristic cerebral oscillatory (about 10/sec.) after-discharges evoked by single cerebral stimulus, within a few minutes after series of medulla stimuli (cf. B_2). Simultaneous recording from medulla explant shows rapid onset of long-lasting, repetitive spike barrage (at high frequency). (C_2): Similar medulla and cerebral after-discharges following paired medulla and cerebral stimuli. Note that medulla spikes tend to occur in bursts which are somewhat synchronized with the positive phases of the cerebral oscillatory sequences. (C_5): Positive cerebral potential, of small amplitude and long duration, evoked by single cerebral stimulus. This may be followed by a positive slow wave of much larger amplitude ($C_{3, 4, 6}$), with a latency of about 200 msec. Note that onset of high-frequency spike barrages in medulla tends to be delayed during interval between onset of early and late cerebral evoked potentials. On the other hand, when cerebral stimulus is ineffective in triggering later cerebral slow wave (C_5), spike barrage in medulla begins shortly after cerebral stimulus and continues unabated for a long period (cf. $C_{3, 4, 6}$). Even when a repetitive spike discharge begins spontaneously in the medulla explant *before* application of a cerebral stimulus (C_4), it appears to become rapidly quenched and does not develop again until later cerebral slow wave occurs.

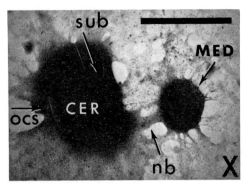

Fig. 7 (X)

records from the medulla after a local cerebral stimulus (Fig. $7C_{3, 4, 5}$), for example, may be due merely to variations in the latency of triggering medulla spike activity, but the sudden onset of the latter activity at the same time as the appearance of a large positive cerebral slow-wave response is most intriguing (see also Fig. $8B_5$). Of further interest in this regard is the observation that a less effective cerebral stimulus—one which failed to trigger a long-lasting cerebral after-discharge—nevertheless evoked a sustained spike barrage in the medulla after a much *shorter* latency (compare Fig. $7C_5$ and Fig. $7C_{3, 4, 6}$). Even when a repetitive spike discharge began spontaneously in the medulla explant *before* the application of a cerebral stimulus (C_4), it appeared to be rapidly quenched and did not develop again until the onset of the cerebral positive slow-wave. These data suggest that local cerebral stimuli can activate neurons which may interfere with generation of spike barrages from the medulla.

Furthermore, spontaneous recordings obtained in this (and another) culture (after return from strychnine to control medium) showed rhythmic series of *positive* cerebral slow-waves arising shortly after each prominent spike burst in the medulla explant (Fig. 9A–F). It has been suggested previously that the positive and negative slow-waves generated by these cerebral explants represent extracellular records of summated inhibitory and excitatory post-synaptic potentials, respectively (Crain and Bornstein, 1964; Crain, 1966). The present data might then be interpreted as indicating the periodic generation of impulses in medulla "pacemaker" neurons which, in addition to producing cerebral excitatory effects, also propagate through inhibitory neurites to cerebral neurons, where they generate large inhibitory post-synaptic potentials. The medulla spike barrage may then be periodically self-quenched (possibly by local recurrent inhibitory networks), thereby attenuating the inhibitory activity to the cerebral explant.

In some cases, these repetitive sequences of spike bursts from the medulla alternating with silent periods have been seen even when cerebral activity disappeared for various intervals. In other cases, however, spike barrages from the medulla appeared to become longer-lasting and less sharply interrupted by silent periods when the cerebral explant became quiescent (Fig. 9G). The latter observation raises the possibility that inhibitory feedback from the cerebral to the medulla explant may also play a role, under some conditions, in quenching medulla spike activity (Fig. 9A–F: note negative cerebral waves concomitant with silent periods in medulla). In

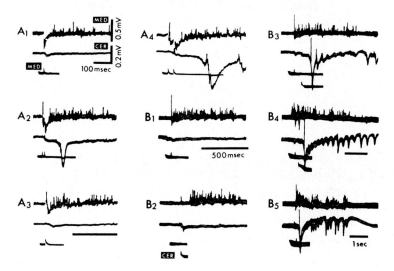

FIG. 8. Temporal and spatial facilitation in coupled cerebrum–medulla explants (same culture as in Fig. 7).
(A_1): Single stimulus to medulla evokes long-lasting spike barrage in nearby region of same explant, but only a small, early-latency response appears in cerebral explant. (A_2): Temporal facilitation occurs with paired stimuli to medulla (at 12 msec. test interval), leading to triggering of large cerebral slow wave after about 100 msec. latency. ($A_{3, 4}$): Similar sequence recorded at faster sweep rate. Note development of cerebral oscillatory after-discharge following paired stimuli to medulla (A_4). (B_1): Single medulla stimulus fails to trigger cerebral after-discharge (as in $A_{1, 3}$), although medulla spike barrage continues for longer than 500 msec. (B_2): Small cerebral stimulus elicits only a brief response, in neighbouring cerebral tissue, even though a spike barrage of long duration develops in the medulla. (B_3): Spatial facilitation occurs with pairing of these medulla and cerebral stimuli (even when latter follows former by intervals as long as 75 msec.), leading to triggering of complex cerebral oscillatory after-discharge. ($B_{4, 5}$): Similar facilitatory effects as in B_3, recorded at slower sweep rates. Note, in B_4, that medulla spike barrage had already begun spontaneously before the medulla explant was stimulated. Also note rhythmic (about 2/sec.) changes in amplitude of positive waves in cerebral oscillatory after-discharge (B_5) and the concomitant synchrony in the medulla spike-burst pattern (see also Fig. $7C_{1, 2}$ and Fig. 9).

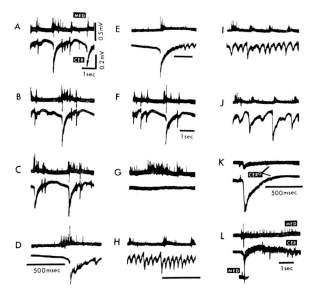

Fig. 9. Excitatory and "inhibitory" interactions between coupled cerebrum–medulla explants during spontaneous activity (same culture as in Fig. 7).

(A–F): Rhythmic discharges occurring synchronously between medulla and cerebrum explants (after return from strychnine to control medium). Note that large *positive* cerebral slow wave tends to occur regularly after each prominent spike burst in medulla. This is seen more clearly at faster sweep rate (D), but at times cerebral discharge may lead that in medulla (E; see also Fig. $7C_{2, 3, 4, 6}$). Also note correlation of silent periods in medulla spike-burst patterns with occurrence of *negative* cerebral slow waves. (G): During intervals when spontaneous cerebral activity is absent, spike bursts in medulla appear to be longer-lasting and less sharply interrupted by silent periods. (In other cases, however, repetitive sequences of medulla spike bursts alternate with silent periods even in *absence* of cerebral activity—see text.) (H–J): Development of *sustained* oscillatory discharges in cerebrum concomitant with appearance of more rhythmic (about 1–2/sec.) spike bursts in medulla—in contrast to the more typical occurrence of periodic "damping" of cerebral oscillatory discharges (as in A–F; also note transition back to latter state in J). (K): Large positive evoked potentials recorded from two sites in another cerebral explant (CER°) in the same culture, but *not* coupled to a medulla explant. No signs of oscillatory after-discharges or spontaneous activity could be detected in this isolated explant. (L): After transfer of electrodes back to coupled cerebrum–medulla explants, long-lasting cerebral oscillatory after-discharge is still evoked by a single stimulus to medulla (and the complex discharges also continue to occur spontaneously).

addition, excitatory influences of cerebral neurons on the medulla are probably also involved, since the onset of discharges in the former precedes, at times, onset in the latter (Fig. 9E; see also Fig. $7C_{2,3,4,6}$). These remarks are, of course, highly speculative, but they provide a working hypothesis for further experiments on these complex heterogeneous neuronal arrays, incorporating microelectrode recordings at multiple sites within each explant and correlative intracellular measurements.

Sustained EEG-like activity has not been generally observed in our isolated cerebral explants (Crain, 1966). It is therefore interesting to note the appearance of "undamped" oscillatory (about 10/sec.) activity in this (and another) coupled cerebral explant *concomitant* with the occurrence of rhythmic (about 1 or 2/sec.) medulla spike bursts (Fig. 9H–J; cf. Fig. 9A–F). Cerebral excitability may have been sufficiently increased during these periods so that 1/sec. bursts of impulses from the medulla explant are adequate now to prevent the cerebral oscillatory discharges from rapidly attenuating within the first second, as they often tend to do (e.g. Figs. $7C_2$; $8B_5$; 9A-F; see also Crain, 1966). The generation of these elaborate activities is in sharp contrast to the absence of spontaneous discharges in an isolated ("control") cerebral explant in the same culture. Responses evoked in the latter explant were also much simpler and briefer in duration (Fig. 9K compared to 9L). Similar differences have been seen in two other coupled as contrasted to isolated cerebral explants but the wide variations in the excitability of isolated cerebral explants (Crain, 1966) indicate that more systematic studies are necessary to clarify the degree to which coupling with medulla (or other CNS) tissues may modify intrinsic cerebral functions. It is hoped that these and related experiments will soon be facilitated by the development of sterile culture chambers with sealed-in microelectrodes to permit chronic recordings under steady-state conditions.

DISCUSSION

Immature neurons from different regions of the CNS, e.g. cerebrum, cord, can therefore make functional (presumably synaptic) connexions when they are permitted to grow towards each other on a transparent collagen film in tissue culture. In the present study, the neuron perikarya were maintained within their normal tissue framework (in 1 mm.^{3} explants), and junctions were established by neuritic branches of the nerve cells. Although this method has obscured direct observation of the neuritic pathways and details of formation of connexions, it has nevertheless established the capacity of CNS neurons to retain synaptogenic properties during development in the artificial culture environment. Recent electron microscope studies by Stefanelli and co-workers (1967) have shown that this capacity for synaptic formation in culture may be present even after complete dissociation by trypsin of immature retinal cells (from the four-day chick embryo eye-cup). Further development of enzymic, or other, dissociation techniques may make it possible to carry out correlative bioelectric and cytological studies on much smaller groups of neurons where the terminal branches of each cell, as well as the perikaryon, may be more clearly observed during all stages of differentiation and formation of interneuronal relationships in culture. This would greatly facilitate experimental analysis of the development of specific neuronal connexions *in vitro*. We may find, however, that the types of neuronal organization which can form under isolated conditions in culture will tend towards the "diffuse, network-like systems" emphasized recently by Székely (1966), in which specificity of neuronal connexions may not play a prominent role. With a judicious choice of embryonic CNS and associated tissues, for example from the visual system, neuronal arrays based on highly ordered selective connexions may nevertheless be feasible for direct study in culture as an extension of the elegant work, *in situ*, of Sperry, Gaze, and others (see papers in this volume).

Together with such attempts to simplify the nerve cell assemblies in this model system, it will be of interest to carry out more intensive studies on the mechanisms underlying the orderly series of developments which we now know can occur in our present CNS explants of 1 mm.^{3}. For these neurons (of the order of 1,000 per explant) not only form axo-somatic and axo-dendritic synaptic relationships after explantation from immature ("synapse-free") regions of the CNS, but also become organized into networks capable of generating complex bioelectric activities which may mimic basic patterns of the CNS, *in situ*, including important components of the mammalian EEG. The capacity of the neurons in a CNS explant to form functional connexions with neurons in separate explants of other types of CNS tissues adds still another dimension to the experimental flexibility of this model system. The application of more sophisticated biophysical techniques, including intracellular microelectrode recordings, and improvements in microscopic resolution of cytological structures within our current multilayered explants, offer great promise for experimental analysis of significant cellular mechanisms underlying the organotypic CNS activities which develop in these cultured tissues.

Recent experiments have shown that these complex functions can still develop within a cerebral or cord explant even when the immature (early foetal rodent) tissue is maintained in a state of bioelectric inactivity by the introduction of functional blocking agents (e.g. anaesthetics, high Mg^{++}) during the entire period in culture (Crain, Bornstein and Peterson, 1968). These observations are in agreement with, and extend, the classical *in situ* studies of Harrison (1904), Carmichael (1926), and others, on behavioural development in amphibian embryos, and they demonstrate the potentialities of mammalian CNS cultures for further analyses of the role of genetic programming, as compared to extrinsic factors, during critical stages of CNS differentiation.

Chronic exposure of coupled CNS explants to agents blocking bioelectric activity may also help to clarify some of the mechanisms of interaction between cerebral and brain-stem tissues. If, in the presence of these chemical agents, synapses can still form between neurons of different CNS explants, it should be possible to evaluate the role of trophic factors as compared to afferent input in modifying the function of cerebral tissue. *Selective* depression (or excitation) of the bioelectric activity of particular types of CNS neurons by chronic incorporation of suitable pharmacological agents in the culture medium may provide still another valuable tool for analysis of these interactions.

SUMMARY

Functional interneuronal connexions have been shown to form between explants (1 mm.³) of various foetal central nervous tissues, separated by gaps of about 1 mm., during maturation in culture. Electrophysiological analyses indicate that impulses evoked by electric stimuli in a dorsal root ganglion can trigger complex bioelectric activity in attached spinal cord tissue, leading to long-lasting discharges in *de novo*-coupled brain-stem or cerebral explants. The transmission of activity from one explant to another occurs after long latencies and is probably due to gradual activation of the CNS neurons through polysynaptic pathways within and between explants. Spontaneous discharges often occur sporadically and may be well synchronized between neurally linked explants. Bioelectric interactions between cerebral and medulla explants appear to be particularly complex and suggest that inhibitory as well as excitatory mechanisms may be involved. Preliminary data indicate that the bioelectric properties of cerebral explants may be significantly modified when such explants are coupled to brain-stem tissues.

These experiments demonstrate the potentialities of CNS tissue cultures as a model system to supplement and extend *in situ* studies of interactions among various types of CNS neurons, especially during growth and regeneration.

Acknowledgements

This study was begun by Dr. Crain and Mrs. Peterson while they were in the Departments of Anatomy, Neurology and Surgery, at the College of Physicians and Surgeons, Columbia University, New York, N.Y., during their association with the Laboratory for Cell Physiology (at that institution) which is directed by Dr. Margaret R. Murray. The preliminary work was supported there by grants NB-03814 (to Dr. Crain) and NB-00858 (to Dr. Murray) from the National Institute for Neurological Diseases and Blindness, U.S. Public Health Service (see also footnote on first page of paper).

REFERENCES

BORNSTEIN, M. B. (1964). In *Neurological and Electroencephalographic Studies in Infancy*, pp. 1–11, ed. Kellaway, P., and Petersén, I. New York: Grune & Stratton.
BUNGE, M. B., BUNGE, R. P., and PETERSON, E. R. (1967). *Brain Res., Amst.*, **6**, 728–749.
BUNGE, R. P., BUNGE, M. B., and PETERSON, E. R. (1965). *J. Cell Biol.*, **24**, 163–191.
CARMICHAEL, L. (1926). *Psychol. Rev.*, **33**, 51–58.
CORNER, M. A., and CRAIN, S. M. (1965). *Experientia*, **21**, 422–424.
CRAIN, S. M. (1956). *J. comp. Neurol.*, **104**, 285–330.
CRAIN, S. M. (1964a). *Anat. Rec.*, **148**, 273.
CRAIN, S. M. (1964b). In *Neurological and Electroencephalographic Studies in Infancy*, pp. 12–26, ed. Kellaway, P., and Petersén, I. New York: Grune & Stratton.
CRAIN, S. M. (1965). *Neurology*, **15**, 291.
CRAIN, S. M. (1966). *Int. Rev. Neurobiol.*, **9**, 1–43.
CRAIN, S. M., and BORNSTEIN, M. B. (1964). *Expl Neurol.*, **10**, 425–450.
CRAIN, S. M., BORNSTEIN, M. B., and PETERSON, E. R. (1968). *Acta Univ. Carol.*, in press.
CRAIN, S. M., and PETERSON, E. R. (1964). *J. cell. comp. Physiol.*, **64**, 1–13.
CRAIN, S. M., and PETERSON, E. R. (1965). *Anat. Rec.*, **151**, 340.
CRAIN, S. M., and PETERSON, E. R. (1966). *Fedn Proc. Fedn Am. Socs exp. Biol.*, **25**, 701.
CRAIN, S.M., and PETERSON, E. R. (1967). *Brain Res., Amst.*, **6**, 750–762.
HARRISON, R. G. (1904). *Am. J. Anat.*, **3**, 197–219.
PAPPAS, G. D. (1966). In *Nerve as a Tissue*, pp. 49–87, ed. Rodahl, K., and Issekutz, B., Jr. New York: Hoeber.
PETERSON, E. R., CRAIN, S. M., and MURRAY, M. R. (1965). *Z. Zellforsch. mikrosk. Anat.*, **66**, 130–154.
STEFANELLI, A., ZACCHEI, A. M., CARAVITA, S., CATALDI, A., and IERADI, L. A. (1967). *Experientia*, **23**, 199–200.
SZÉKELY, G. (1966). *Adv. Morphogen.*, **5**, 181–219.

DISCUSSION

Stefanelli: Some results my co-workers and I have obtained (Stefanelli, A., Zacchei, A. M., Caravita, S., Cataldi, A., and Ierida, L. A. [1966]. *Atti Accad. naz. Lincei Rc.*, Ser. VIII, **40**, 758–761; Stefanelli, A., Zacchei, A. M., Caravita, S., Cataldi, A., and Ieradi, L. A. [1967]. *Experientia*, **23**, 199) are very closely linked with Dr. Crain's findings.

The problem of nervous activity has to be investigated at two fundamental levels, the cellular and the supracellular—that is, the single neuron and functional systems of interconnected neurons. Of course it will be easier to understand the activity of a multineuronal system if it has a restricted number of neurons. Many investigators therefore study function in lower animals because these systems are composed of a very few nervous elements, generally of considerable size. Others, like Dr. Crain and myself, try to resolve the problem by culturing *in vitro* only a limited number of neurons.

Both methods can be criticized: on the one hand it is difficult to refer data obtained from low animal classes to more evolved classes, and on the other hand

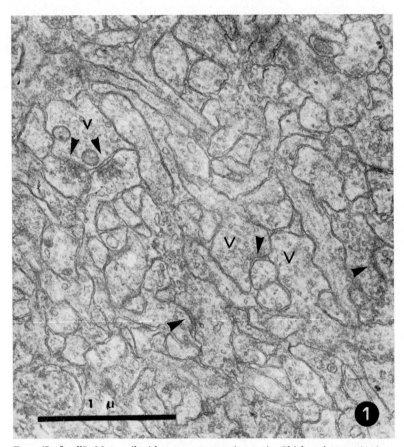

FIG. 1 (Stefanelli). Neuropil with many synapses (arrows). Chick embryo retina incubated for four days and completely dissociated by trypsin, then reaggregated and cultivated *in vitro* for 25 days.

it is difficult to generalize from the results obtained from neurons cultured in an artificial environment. This must be kept in mind when we try to reach a general conclusion.

Dr. Crain, working on tissue explants interconnected by bundles of growing fibres, has obtained new synaptic junctions between different explants. Nevertheless the activity of each explant starts in a pre-formed array of embryonic elements which develop normal interconnexions. We have tried to eliminate these pre-existing interconnexions in embryonic nervous tissue (from four-day-old chick embryos) by dissociating the cells, in order to build *in vitro* a completely new morphological and functional pattern.

The dissociated cells were handled in two ways:
(1) They were cultivated as monolayers with different supporting substrates, their number and the mixture of cell types being varied.
(2) They were allowed to reaggregate in pellets about 0·5 mm. in diameter.
The cells were maintained in culture for various times.

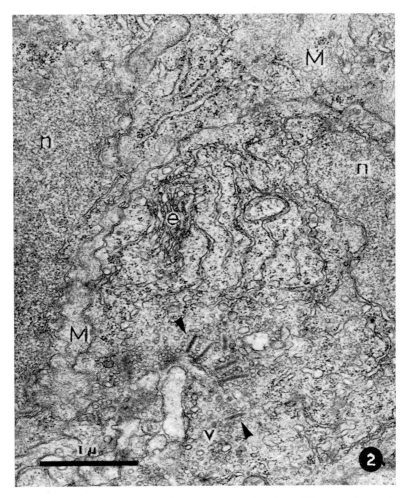

FIG. 2 (Stefanelli). Synaptic pole of a photoreceptor showing a rich array of synaptic ribbons (between arrows) and synaptic vesicles (v). The endoplasmic reticulum rich in ribosomes (e) and the nucleus (n) are indicated; note the Müller cells (M) containing gliofilaments. Same type of preparation as in Fig. 1 (Stefanelli).

With the first method of culture we succeeded in obtaining elementary nervous microsystems. Unfortunately, we were only able to obtain good electron microscope pictures in a few cases. However, we shall continue these studies, and it would be of great interest to test the bioelectric activity of such simple systems.

We have had more successful results with the second method. It was easier to get long-term cultures and in all the dissociated tissues the cells tended to re-aggregate histotypically. Furthermore, it was easier to preserve the reaggregated pellets for electron microscope observations and to find morphological structures such as synapses.

We cultivated various types of reaggregated cells and we also mixed cells of different histological types. Here I shall report the results obtained by reaggre-

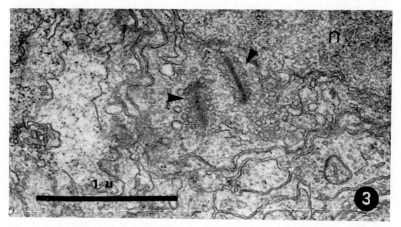

Fig. 3 (Stefanelli). Another example of well-differentiated synaptic ribbons. Same type of preparation as in Fig. 1 (Stefanelli).

gating chick embryo retinal cells dissociated at four days of age and maintained *in vitro* for up to 50 days. For our purpose, the retina is a very interesting nervous tissue because of its variety of sensory and other types of cells, including Müller's cells. Together with the electron microscope observations, I shall use light and electric stimuli to study the bioelectric activity of these cultures. However, at present I can report only on the morphological aspects of the problem.

We found the following:

(1) Autodifferentiation of nerve cells, photoreceptors and Müller's cells *in vitro*;

(2) A tendency of cells to reaggregate histotypically; if present they appear to be arranged in characteristic "rosettes", or groups of pigment epithelium cells facing each other;

(3) Differentiation of a dense neuropil, very rich in synapses which are typical and apparently functional (Fig. 1);

(4) Complete differentiation of the synaptic pole of the photoreceptors with characteristic synaptic ribbons (Figs. 2 and 3);
(5) Various degrees of differentiation of the "comb-like" structure of the outer segment derived from the flagellum (Figs. 4 and 5).

It appears from our reaggregated retinas that the pigment epithelium cells play an important role in the differentiation of the "comb-like" structure of the outer segment of the receptor. In cultures without pigment epithelium cells, the outer segments remain in an early stage of differentiation, as club-like processes containing the filaments of the flagellum and without any tendency to form tubules and flat disks from the cell wall. Furthermore, the histological texture is much less regular.

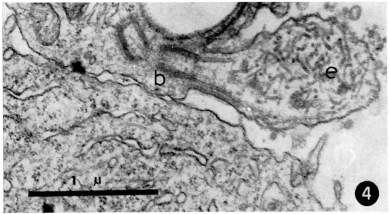

FIG. 4 (Stefanelli). An early stage of development of the outer or external segment (e), showing ciliary filaments and tubules. (b), basal bodies. Same type of preparation as in Fig. 1 (Stefanelli) but cultured for 28 days.

The importance of the pigment epithelium as the substrate in which the 11-*trans* isomer of vitamin A is concentrated and presumably is changed into the chromophore of the visual pigment (11-*cis* retinene) has recently been stressed by Sorsby, Reading and Bunyan (Sorsby, A., Reading, H. W., and Bunyan, I. [1966]. *Nature, Lond.*, **210**, 1011–1015). Moreover it has been demonstrated that the outer segment of the photoreceptor degenerates in mice deficient in vitamin A (Dowling, I. E., and Gibbons, I. R. [1961]. In *The Structure of the Eye*, ed. Smelson, G. K. New York: Academic Press), that in some strains of albino mice the rods do not differentiate (Sorsby, A., Koller, P. C., Attfield, M., Dovey, I. B., and Lucas, D. R. [1954]. *J. exp. Zool.*, **125**, 171–187), and lastly that after treatment with potassium iodide there is degeneration of the pigment epithelium, with consequent degeneration of the outer segment of the rods (Sidman, R. L. [1961]. *Dis. nerv. Syst.*, **32**, monograph suppl.). We may add that Sidman demonstrated the inability of the retina maintained *in vitro* to convert 11-*trans* vitamin A into

11-*cis* retinene and also to use 11-*cis* retinene supplied with the culture medium. This is probably due to the lack of opsin, the protein normally bound to 11-*cis* retinene to form rhodopsin.

These results clearly indicate the importance of the pigment epithelium as an organizer of the differentiation of the sensory pole of photoreceptors.

But it is most significant that the synaptic apparatus is perfectly differentiated in spite of the lack of differentiation of the outer segment. That is to say, there is no functional causation of synaptic differentiation. This must be kept in mind when we wish to assess the contribution of function as opposed to structure during embryonic development. Histogenetic determination is operative even down to the level of fine structure.

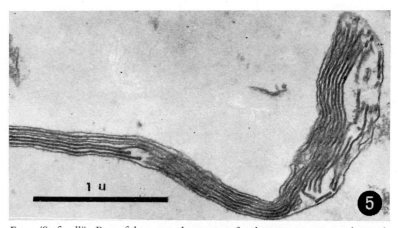

Fig. 5 (Stefanelli). Part of the external segment of a photoreceptor at an advanced stage of differentiation, showing its characteristic lamellae. In this case even the pigment cells of the tapetum were present in the aggregate and they seem to be necessary for the differentiation of the lamellae. Same type of preparation as in Fig. 4 (Stefanelli).

Kerkut: Dr. Crain, would raising the calcium level of your medium stop the oscillatory response in the neurons?

Crain: Raising the magnesium level blocks the oscillatory responses as well as the primary evoked potentials, but a fivefold increase in calcium concentration has not produced any significant interference with these response patterns. These data support the view that the oscillatory after-discharges in cerebral explants are mediated by synaptic circuits and are not merely due to oscillatory properties of the individual nerve cell membranes (Crain, 1966, *loc. cit.*).

Hník: I believe you suggested, Dr. Crain, that the increased excitability you observed might be due to the formation of a more extensive synaptic network in your tissue cultures. Could it not, however, be due to quite the reverse mechanism? As has been shown by W. B. Cannon and A. Rosenblueth (1949. *The Supersensitivity of Denervated Structures: A Law of Denervation.* New York:

Macmillan Co.) and by C. G. Drake and G. W. Stavraky (1948. *J. Neurophysiol.*, **11**, 229–238), increased sensitivity to mediators and to nerve impulses may be present when less of the surface of the neuron is covered with synapses.

Crain: Undoubtedly there are many abnormalities and pathological patterns in these isolated CNS tissues, and the hyperexcitability that we see may indeed reflect deficits in synaptic components. My remarks on the hyperexcitability and complexity of the response patterns in these cultures were intended to emphasize that at least a moderate degree of synaptic organization underlies these bioelectric activities. In cultures of neurons devoid of synapses, e.g. dorsal root ganglia or excessively dissociated CNS tissues, no such long-lasting, complex discharges have been detected—only simple, brief spike potentials are evoked (Crain and Peterson, 1964, *loc. cit.*). My point is that synaptic networks are still being maintained in these CNS explants, with properties resembling—but certainly not identical to—their *in situ* counterparts. D. P. Purpura and E. M. Housepian (1961. *Expl Neurol.*, **4**, 377–401) suggested that the hyperexcitability of chronically isolated slabs of neonatal cerebral cortex, *in situ*, is due to extensive axon-collateral sprouting that occurs during the regeneration of pyramidal neurons after surgical isolation, leading to increased excitatory synaptic connexions. This is only one of several possible explanations of the hyperexcitability. I agree that other phenomena may also be involved, relating to synaptic deficit, abnormal ratios of excitatory and inhibitory synapses, and hyperexcitability of the denervated neuronal membranes (Crain, 1966, *loc. cit.*).

Székely: Is there any way to decide which neural structures were stimulated when you applied electrical stimulation to your explants? H. Grundfest (1959. In *Handbook of Physiology: Neurophysiology*, Vol. I, pp. 147–197, ed. Field, J., Magoun, H.W., and Hall, V. E. Washington, D.C.: American Physiological Society) maintains that since the membrane of the perikaryon is completely covered by synaptic junctions, it is electrically inexcitable. Is it possible to bring the electrode quite close to a nerve cell under the control of a microscope and to investigate the response to electrical stimulation in young explants in which the nerve cell is not yet covered by synapses, and in older explants in which the synapses have already developed?

Eccles: I think everybody now believes that the perikaryal membrane is electrically excitable. The precise areas covered by synapses are a different story, but no one has investigated that. Actually at the amphibian neuromuscular junction, B. Katz and R. Miledi (1965. *Proc. R. Soc. B.*, **161**, 453–482) have shown that the impulse propagates to the very terminals of the nerve fibres lying in the grooves on the muscle surface.

Crain: The data from cultured neural tissues are in complete agreement with what you say, Professor Eccles. W. Hild and I. Tasaki (1962. *J. Neurophysiol.*, **25**, 277–304) have shown, for example, that dendrites located more than 100 μ away from the perikaryon of cerebellar neurons can be electrically excited and these impulses can then propagate to the cell body. I have seen similar phenomena in cerebral cortex cultures, where neurites that grow out for several hundred μ from the pial edge of the explant can be excited peripherally, leading to propagation of impulses into the explant. In none of these experiments with cultured tissues, however, was there any evidence of synaptic junctions on the neuronal membranes where impulse propagation occurred. Critical experiments will require the preparation of cultures containing dendrites which are not only sufficiently visible in the living state to permit selective microelectrode placements, but which are also in synaptic contact with large numbers of axons. Unfortunately, we have obtained bioelectric evidence of synaptic activity only from relatively dense regions of CNS explants, where neuronal surface contours and synaptic junctions have not been resolvable in the living cultures because of their poor contrast with surrounding glia. Possibly with interference microscopy we shall be able to visualize these structures.

Murray: Have you tried the Nomarski system?

Crain: Preliminary observations with Nomarski interference optics appear promising.

Gaze: A combination of your approach, Dr. Crain, and that described by Professor Stefanelli would be interesting. Your work already involves considerable technical virtuosity, but may I suggest something to make the situation even more difficult? Could the direction of the outgrowing fibres be controlled by a technique such as that recently described by S. B. Carter (1965. *Nature, Lond.*, **208**, 1183–1187), in which cellulose acetate was shadowed with metallic palladium so that cells would only grow on one surface and not on the other? It appears that the direction of cell migration can be controlled in this way. Whether this can be done for the direction of axonal growth, we don't yet know. If it can, one could make a sort of biological computer by having channels specially constructed for the individual axons. With any luck one might be able to direct the formation of connexions.

Crain: This is an extremely interesting approach. Some of Weiss's earlier studies of contact guidance of neurites growing in tissue cultures (1945. *J. exp. Zool.*, **100**, 353–386), might be fruitfully extended by use of these finely controllable substrates.

With regard to Professor Stefanelli's studies of dissociated cells, we have also attempted to dissociate neural tissues, but only peripheral ganglion cells have regenerated well in monolayer cultures. This is why we have limited our present studies of synaptic networks to conventional explants, pending further development of dissociation techniques for CNS tissues. Professor Stefanelli's finding that synapses can indeed form after explantation of aggregates of completely dissociated retinal cells lends additional hope to the possibility that these enzymic methods may permit us to prepare more orderly arrays of CNS cells, in which we can visualize both intercellular and intracellular details during the entire course of development. Professor Stefanelli, when do the first synaptic junctions appear in these aggregates? Presumably there are none at explantation.

Secondly, if the dissociated retinal cells are laid out in culture as a single layer rather than as a three-dimensional pellet, do any synaptic junctions form?

Stefanelli: We have not yet been able to carry out monolayer cultivations of single cells for a long period. We have light microscopic evidence of some contact, but we have no electron microscopic proof that real synaptic junctions exist. For the moment we have left this field, because we must first find a better way to cultivate the single cells. At birth in the chick embryos some synapses are present, but at 21 days there are none in single layer cultures of dissociated retinal cells. With the method of reaggregation, synapses appear within 13 days of culture.

Sperry: You mentioned something about the lack of specificity in your cultures, Dr. Crain. We should remember here that even the transplantation of a clump of nerve cells into a foreign part of the body may be sufficient to destroy the fine specificities that are involved in the functional hook-ups. If one wants to study specific connexions it might be better either to go back to something like Speidel's old method, utilizing the transparent tadpole tail, or to implant chambers in the body itself.

Crain: Although specificity has not been detected in our present studies with CNS explants, cultures of more favourable tissues, e.g. retina and optic tectum, may provide a basis for analysis of the minimal cellular organization prerequisite for development of the highly specific functional connexions characteristic of many parts of the CNS *in situ*. [Note added in proof.]

Székely: In some of your diagrams, Dr. Crain, very nice rhythmic activity appeared. Did you see any difference in histological structure between the explants which produced these bursts and those which did not? Was there a difference in the number of neurons per unit area between the two types of explant?

Crain: As yet we have no significant correlations between the histological patterns in the cultures which show regular rhythmic activity and those which are relatively quiescent unless stimulated. Histology of the explants, both at the light microscope level and the electron microscope level, shows that in all of these cases there are abundant synapses, but there are no clear-cut differences between the two types.

Eccles: No one has mentioned the question of inhibition. Maybe a lot of the excessively convulsive activities of Dr. Crain's explants could be due to the fact that the inhibitory mechanisms did not grow adequately.

Crain: Emphasis has been placed on the more complex "convulsive" activities of the CNS explants to dramatize some of the organotypic functional capacity of these isolated tissues. As noted in my presentation, however, the cerebral and cord explants generally show relatively little spontaneous activity in normal culture media. Since strychnine leads to increased spontaneous activity, as well as to longer-lasting, more complex, electrically evoked after-discharges, inhibitory mechanisms are probably involved in maintaining the relatively quiescent state under normal culture conditions (see also Crain, 1966, *loc. cit.*). [Note added in proof.]

Sperry: Professor Szentágothai emphasized how far we had already come 18 years ago at the Chicago conference on genetic neurology. This reference is appropriate inasmuch as six of us here at this symposium were also at that earlier conference. However, in using this 18-year measuring stick, we leave out the highly critical decade 1939–1949 during which most of the radical revisions in our thinking that Professor Szentágothai referred to had already taken place. The first five years were particularly critical; by 1945 (see Sperry, R. W. [1945]. *J. Neurophysiol.*, **8**, 15–28; [1965]. In *Organogenesis*, ch. 6, ed. De Haan, R. L., and Ursprung, H. New York: Holt, Rinehart and Winston) most of the basic concepts in the developmental patterning of synaptic connexions, as we see them today, had already been spelled out in the literature.

REGULATIVE FACTORS IN THE ORDERLY GROWTH OF RETINO-TECTAL CONNEXIONS

R. W. SPERRY and E. HIBBARD

*Division of Biology, California Institute of Technology,
Pasadena, California*

As a model for the study of the developmental patterning of brain pathways and connexions, the optic tract of lower vertebrates connecting the retina of the eye to the midbrain tectum offers a number of long-recognized advantages. Much of our current understanding of the developmental organization of brain circuits stems from experiments on this fairly simple and accessible optic system. There is good reason to think that the basic developmental mechanisms found to operate in the prefunctional shaping of this system for visual behaviour have wide application to the developmental patterning of the nervous system in general.

Strange as it may seem today, the first experiments on the growth of behavioural organization in this system were designed to determine whether the basic visual properties, like perception of directionality, size, movement, contour, and so on, are installed directly by the growth process itself or have to be organized through use and behavioural adjustment (Sperry, 1965). This was a little over 25 years ago now, when it was still believed that the growth of nerve fibres was non-selective, subject only to mechanical guidance. Chemical and electrical guidance appeared to have been ruled out, and several lines of evidence seemed to show that selective synaptic connexions are unimportant, at least for orderly behaviour. "Instinct" was still a bad word in science and the whole concept of the inheritance of behavioural patterns remained anathema to prevailing doctrines in psychology. No developmental machinery was then known or available, even in theory, by which the highly ordered and precisely designed neural connexions for behaviour could be "grown into" the nervous system without the aid of function. Neurobiotaxis, with its dependence on function for selectivity, was still the prevailing favourite among explanatory concepts. Even the old master, Cajal, had been willing to leave it to function when it came to the detailed central adjustments of synaptic associations for behaviour.

We have come a long way since then, of course, and the issue of growth versus function seems now to be quite settled and dead, at least in the case of retino-tectal connexions. The question is nevertheless of some current concern with respect to many of the other fibre tracts of the brain, particularly in man. It would seem fairly safe to speculate in this regard that the patterning of most of the long fibre systems of the CNS, even in man, is primarily a problem of developmental mechanics, not one of learning. Another way of saying the same thing is that, if learning involves changes in the morphology of nerve connexions at all, these would appear to be confined to dendritic relations and local neuron circuits without affecting very much the patterning of the long fibre systems of the brain.

Once the growth-versus-function issue was settled in favour of growth (Sperry, 1944), attention was turned to the nature of possible developmental mechanisms. The next general question—and one still pertinent and basic to the nervous system as a whole—concerned the extent to which orderly selective function is dependent on selectivity and orderliness in the underlying neural connexions. As late as 1954 it was being suggested that orderly visual recovery might be explained even though the divided optic fibres reconnected in a random diffuse pattern, with retino-tectal communication based on a physiological coding–decoding scheme. Similarly the pre–1940 Resonance Principle of Weiss (1937) and a more recent interpretation along similar lines proposed by Szentágothai (1961) and Székely (1966) would provide for selective communication regardless of absence of selectivity in the morphology of the fibre connexions.

Evidence that the normal topographical projection on the tectum is in fact restored after section and scrambling of the optic fibres came first from experiments in which localized lesions were made in the frog tectum. The position and size of the resultant scotoma or blind area were found to be the same whether the localized lesion was made before division of the optic nerve or after regeneration (Sperry, 1944)—a result obtainable only if the regenerated projection map had been restored in its original pattern. Further confirmation on a more refined scale has since come from electrical mapping methods, particularly in the studies of Gaze and his associates (reviewed in Gaze, 1967). Confirmation has also been obtained with histological procedures (Attardi and Sperry, 1960, 1963). The patterning of the optic fibre pathways and their terminal connexions was shown in the latter studies to be selectively predetermined according to the locus of origin of the optic fibre in the retinal field. Fibres from different retinal sectors were found to exhibit different growth patterns, like different species of plants, by which they selectively regained their various predetermined target zones in the tectal field.

Other experiments, with H. L. Arora and A. J. Limpo (see Sperry, 1965), have shown that colour perception—like position, directionality, movement, size, and contour perception—undergoes an orderly reinstatement in optic nerve regeneration. In summary, all the basic functional attributes of vision appear thus to be restored in regeneration of the scrambled optic fibres. Endogenous physiological properties of different sensory and neural cell types were presumed to be involved, as well as synaptic connexions. Where the experiments have been extended into prefunctional stages involving initial nerve outgrowth, similar results have been obtained (see Székely, 1966).

As early as 1942 it was inferred (Sperry, 1942) that optic fibres arising from different retinal loci must be distinguished from one another according to the location of their ganglion cell bodies in the retinal field, "probably through differential physical-chemical properties induced in them by differentiation of the optic cup in development". It was suggested further (Sperry, 1945) that the retina must undergo a gradient or field type of differentiation along at least two separate axes and that contralateral eye transplants might demonstrate that the anteroposterior gradient is established before the dorsoventral gradient. A corresponding biochemical specification of the tectal field was likewise inferred and it was postulated that differential affinities must exist between the tectal neurons and the optic fibres. With supporting results from the vestibular, cutaneous, central and motor systems (Sperry, 1951a, b) the basic concepts were extended to the development of the nervous system generally. This material has been reviewed in some detail on four separate occasions in the past three years (Gaze, 1967; Jacobson, 1966; Sperry, 1965; Székely, 1966) and does not need further amplification.

Recent work on the optic system has been aimed at more detailed analysis of the various phases of the postulated specification processes, chemical-affinity mechanisms, and related factors. Gaze, Jacobson and Székely (1963) tested the elasticity of the tectal gradients in response to ingrowth of fibres from compound eyes composed of two nasal or two temporal half-fields. In this interesting experiment it appeared as if the recipient tectal half-field had spread to cover the entire tectum, producing a rather radical change in the pattern of tectal differentiation. However, as we suggested to Dr. Székely in a seminar at the Massachusetts Institute of Technology a few years ago (see also Sperry, 1965), one should probably reserve judgment on the interpretation of these results until the effects of tectal hyperplasia and hypoplasia, in response to the overload and diminished innervation, can be ruled out. The spread of the innervated half of the tectal gradient into the denervated half may be more apparent than real and more mechanical than chemical. Related problems are mainly of embryological interest. They bear primarily on the developmental dynamics of the morphogenetic field and have only remote significance for problems of brain function.

A question with more direct neurological implications concerns the degree of local resolution and overlap in the optic fibre projection on the tectum. How large is the terminal arborization field of a single fibre and with how many tectal cells over how large a dendritic field does one optic fibre synapse? How much overlap is there in the synaptic zone of neighbouring individual fibres? How do these compare in the normal and in the regenerated condition? The answers to these and related questions may be expected to differ in different species and for different parts of the retina. At this stage a rough approximation would be helpful in judging the higher degrees of precision required in the regeneration process to restore normal vision. The maps based on electrical procedures are difficult to interpret in this connexion because of the variables associated with changes in amplification and the uncertainty whether the evoked potentials come from optic fibre terminals or from post-synaptic firing in the tectal neurons.

Using Golgi staining of the normal two-day-old chick tectum we have seen terminal arborizations of the optic fibres in considerable numbers, measuring about 75 μ or approximately 1/200th of the total arc of the central tectal field. A somewhat lower order of refinement is probably present in the cichlid fishes like the angel fish *Pterophyllum scalare* and *Astronotus ocellatus*, in which we have seen visual recovery with little if any loss in visual acuity. A large—22 per cent—loss of acuity on the average was reported in another study on *A. ocellatus* by Weiler (1966). The extent to which the regenerative process is able to reproduce the normal degree of resolution in visual function needs further study. It is difficult to judge the degree of selective precision of the growth of optic fibres from the pattern established normally because of the complex sequence of possible sustaining factors present in normal development. Observations on the interposition of obstacles and other types of interruptions in the fibre pattern within the tectum (see below) indicate that considerable correction is possible in regrowth within the plexiform layer close to the target zone.

Some years ago one of us (E. H.) noted a very striking similarity in the layered differentiation of retina and tectum in the developing frog tadpole. This is illustrated in Fig. 1 for the ten-week-old tadpole of *Rana pipiens*. While the character of the layering and the thickness of the retina remains fairly uniform throughout, that of the tectum varies somewhat from the

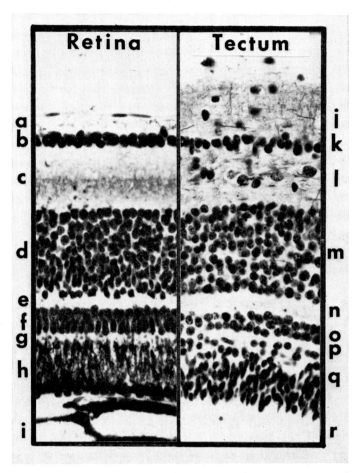

FIG. 1. Comparison of retinal and tectal stratification in a ten-week-old tadpole of *Rana pipiens*. × 400

(a) optic fibre layer
(b) ganglion cell layer
(c) inner plexiform layer
(d) inner nuclear layer
(e) outer plexiform layer
(f) outer nuclear layer
(g) limitans
(h) photoreceptor and pigment cell layers
(i) choroid vessels
(j) marginal and optic layers
(k) superficial grey layer
(l) deep medullary layer
(m) central grey layer
(n) deep white layer
(o) periventricular grey layer
(p) periventricular fibrous layer
(q) subependymal granular and ependymal layer
(r) ventricle

dorsomedial to the ventrolateral portions of the lobe, primarily in the thickness of the periventricular cell layer. The segment shown in Fig. 1 represents the central part of the lobe, corresponding approximately to the position where the centre of the visual field would be projected. The photographs are shown at identical magnifications, the distance from the innermost to the outermost uniform cell layers being 175 μ in both cases. The relative position of the optic nerve fibre layer is the same in both retina and tectum, but the thickness of the tectum is somewhat greater because of the marginal layer overlying the optic fibres. Although the number of cells in each layer arranged perpendicular to the surface is approximately the same (one in the ganglion cell and superficial grey matter, eight to ten in the inner nuclear and central grey matter, two in the outer nuclear and periventricular grey matter), the surface area of the tectum is less than 60 per cent of that of the retina. A considerable amount of convergence between retinal ganglion cells and terminal tectal sites seems indicated. The numbers and positions of cell layers, the distribution and thickness of plexiform layers, and the relative position of outgoing and incoming optic fibres in these structures appear to be more than coincidental.

We have been investigating in a preliminary way, with Dr. Arora (reviewed in Sperry, 1965), the forces that control the growth of optic fibres in various parts of the afferent course into the tectum, using surgical deflections, interruptions and transplantations. When the medial and lateral brachia of the optic tract are cut short and crossed at the anterior tectal pole the bulk of the sprouting fibres promptly turn back to regain their normal channel. When the lateral tract, cut long, is transposed deep into the medial pathway, most fibres enter the medial tract and follow it on across the equator of the tectum into the ventral quadrants before they enter the plexiform layer to synapse in their original zone. When contralateral fibres are superimposed on a fully innervated optic tract they form a plexiform zone only in the corresponding mirror locus of the opposite tectum.

The ability of the optic fibres to find their way to their proper tectal stations even from grossly abnormal starting positions has been investigated by surgical deflection of the cut optic nerve. The central end of the divided optic nerve was apposed to the root of the oculomotor nerve in such a way that the regenerating optic fibres then entered the brain below and caudal to the tectum, near the point where the oculomotor fibres leave. Gaze (1959) had previously reported a case in which the optic nerve inadvertently entered the brain along this route. He was able to record a few evoked potentials on the contralateral tectum corresponding to points of stimulation on the retina from which the nerve had originated, but was unable to elicit optokinetic responses through this nerve. In the present series, the optic nerve was transposed caudally to a point near the roots of the oculomotor and trigeminal nerves in 14 *Xenopus laevis* tadpoles. In three cases, both optic nerves entered the brain through the oculomotor root, and most of the optic fibres succeeded in making their way back to the tectum.

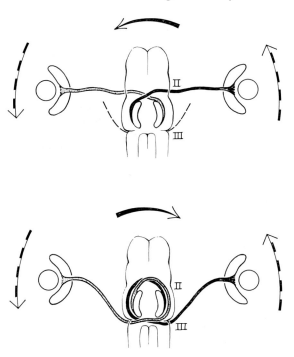

FIG. 2. Diagrams illustrating the courses taken by optic fibres from the eye to the tectum in normal *Xenopus* tadpole (upper), and tadpole in which the optic nerves have been diverted into the oculomotor nerve roots (lower). The response of the normal animal to counter-clockwise rotation of an optokinetic drum, indicated by banded arrows, is to turn to the left. The response of the experimental animal, in which optic fibres cross twice and terminate in the ipsilateral tectal lobes, is to turn in the opposite direction.

However, these connexions were made with the ipsilateral instead of the contralateral tectum—not by following the oculomotor fibres back towards the nucleus of cranial nerve III as might have been expected, but rather by decussating twice, once at the level of the IIIrd root and again in the region of the optic chiasma (see Fig. 2). Behaviourally, this led to reversed optokinetic responses in both directions in two animals, and spontaneous circling to the right in the third (Hibbard, 1967).

In two other cases, the left optic nerve regenerated over the surface of the brain above the meninges. In one of these latter, the aberrant optic nerve was cut again above the midbrain and poked down through a hole into the dorsal surface of the ipsilateral tectum. Thirteen days later this animal began to show spontaneous circling to the left at a rate of about 12 turns per minute. The deflected optic nerves in seven other cases strayed

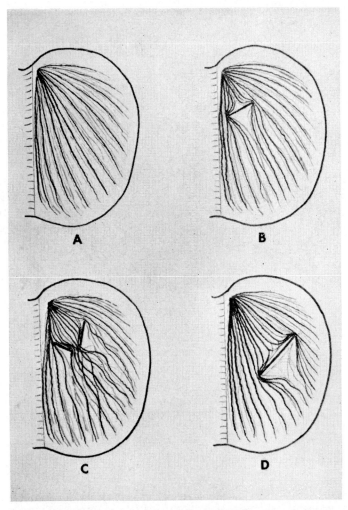

FIG. 3. Routes taken by regenerating optic fibres from the medial tract in the tectum when blocked by tantalum foil barriers. (A) Normal. (B) Small obstacle with fibre-free shadow behind it. (C) Two barriers through which fibres are funnelled. (D) Large barrier with fine fibres in the shadow area behind it.

caudally and entered the brain along the trigeminal nerve root. In traversing the ophthalmic ramus, ganglion, and root of the Vth nerve the optic fibres in these cases remained fasciculated and did not intermingle with trigeminal fibres. The optic fibres could not be followed any distance into the brain within the trigeminal tract and there was no evidence of visual recovery under these conditions.

The orderly, near-parallel distribution of optic fibres in the "parallel" or "radiation" layer of the tectum has yet to be explained. What are the invisible guidelines or forces that keep the fibres from crossing and interspersing and from curling up or dipping down? Are the invisible guidelines primarily mechanical or chemical or a combination of both? In an effort to get some answers to such questions we have made various kinds of experimental interruptions and derangements of the parallel layer in adult fishes, some examples of which are shown in Fig. 3. This work is still in progress, but a few points seem to be emerging. Although the deflected fibres seem not to be able to effect much change of course within the parallel layer itself, the same fibres may show extensive readjustment in the plexiform layer just below. This would appear to permit a considerable degree of target correction near the terminations and to reduce the degree of precision required at choice points in more distal portions of the optic fibre channels. The deflected fascicles undulate through the plexiform layer in smooth curves and gradually straighten out. We have not been able to substantiate Leghissa's (1955) description of a criss-cross overlap among the optic fibres of the parallel layer at the posterior equator.

If, after the optic fibres in the parallel fascicles have been cut, barriers of tantalum foil are inserted across their course, the regenerating fibres pass around the block in a smooth flow, leaving a "shadow" area in the parallel layer relatively clear of fibres immediately behind the block. If two strips of foil are placed at angles of 45° to the parallel bundles in such a way as to form a funnel through which the regenerating fibres will pass, the deflected fibres pass through the narrow opening, at angles of 90° and less, later crossing and re-crossing other fibres largely below or sometimes above the parallel layer proper. They gradually refill the parallel fascicles distal to the block, but whether the fibres are regaining the same channels as they would normally occupy or are merely refilling any empty pathways is not clear. The undulating pattern suggests an inertial component in the fibre growth, with oscillations and corrections of deviations from course by negative feedback.

Correction of the course of the outgrowing axon, apparently in response to an anteroposterior gradient effect, has been clearly shown in abnormally positioned Mauthner cell axons (Hibbard, 1965). If the segment of hindbrain containing the Mauthner cell is reversed at an embryonic stage before the outgrowth of the axon, the axon first grows forward, towards the original posterior end of the grafted segment, but then re-curves to proceed down the length of the spinal cord. In embryos joined back-to-back near the end of the hindbrain (see Fig. 4) the Mauthner axon proceeds along its typical course into the twin, but travelling in the caudo-rostral direction directly opposite to normal. Presumably the microfilament flare at the growing tip does not span sufficient rostrocaudal differences to

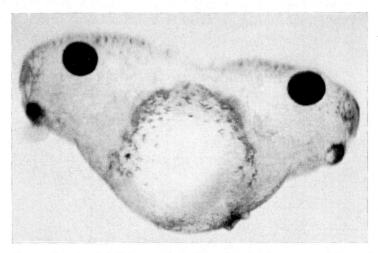

FIG. 4. Two-headed *Xenopus* embryo formed by joining anterior halves of two embryos to study the effect on the course of Mauthner's axons.

overcome the inertial growth component (Swisher and Hibbard, 1967). The direction of advance of a single growing fibre is probably better viewed as a reflection of the algebraic sum of the combined adhesive forces of all the active microfilaments than as the sequence of single strong filament successes. Van der Loos (1965), studying abnormally oriented pyramidal cells in the cortex, has suggested that the initial outgrowth of the axon, in some instances at least, as well as the spread of the dendrites, is intrinsically determined within the cell and is dependent upon the position of the cell, while further outgrowth of the axon is determined by extrinsic factors.

SUMMARY

The nerve tract connecting eye to brain, together with its afferent and efferent association within the retina at one end and within the midbrain tectum at the other, may be regarded as a model system for the study of the developmental organization of brain pathways and synaptic connexions. The physiological and behavioural properties of this system indicate a high degree of complexity, precision, diversity and refinement in the underlying neural relations; any understanding of the regulative factors by which this optic system is organized in growth should therefore take us a long way towards a general understanding of the basic developmental mechanisms for the inherent patterning of brain pathways and connexions.

The orderly formation of retino-tectal connexions, both in development and in regeneration, has been studied extensively in experiments involving surgical, behavioural, histological, electrical, biochemical and related methods, applied separately and in combination. The principal regulative factors disclosed so far by the accumulated evidence have been outlined and some of the behavioural and histological results briefly illustrated. Some newer observations have been presented that deal with the patterning of fibre routes and connexions within the tectum, with a correspondence in retinal and tectal stratification, with the growth patterns of transposed optic nerves, and with the oriented advance of individual axons. An attempt is made to show how the developmental mechanisms inferred from experiments on the retino-tectal relations of lower vertebrates could also account in principle for the ontogenetic organization of the much more complicated mammalian visual system.

Acknowledgements

This work was supported by a grant from the National Institute of Mental Health, MH3372, and by the Hixon Fund of the California Institute of Technology. The technical assistance of Ruth T. Johnson is gratefully acknowledged.

REFERENCES

ATTARDI, D. G., and SPERRY, R. W. (1960). *Physiologist, Wash.*, **3**, 12.
ATTARDI, D. G., and SPERRY, R. W. (1963). *Expl Neurol.*, **7**, 46–64.
GAZE, R. M. (1959). *J. Physiol., Lond.*, **146**, 40.
GAZE, R. M. (1967). *A. Rev. Physiol.*, **29**, 59.
GAZE, R. M., JACOBSON, M., and SZÉKELY, G. (1963). *J. Physiol., Lond.*, **165**, 484–499.
HIBBARD, E. (1965). *Expl Neurol.*, **13**, 289–301.
HIBBARD, E. (1967). *Expl Neurol.*, **19**, 350–356.
JACOBSON, M. (1966). In *Current Status of some Major Problems in Developmental Biology* (25th Symp. Soc. Devel. Biol.), pp. 339–383, ed. Locke, M. New York: Academic Press.
LEGHISSA, S. (1955). *Z. Anat. EntwGesch.*, **118**, 427–463.
SPERRY, R. W. (1942). *Anat. Rec.*, **84**, 470.
SPERRY, R. W. (1944). *J. Neurophysiol.*, **7**, 57–69.
SPERRY, R. W. (1945). *J. Neurophysiol.*, **8**, 15–28.
SPERRY, R. W. (1951*a*). In *Handbook of Experimental Psychology*, pp. 236–280, ed. Stevens, S. S. New York: Wiley.
SPERRY, R. W. (1951*b*). *Growth* **15**, Suppl., 63–87.

SPERRY, R. W. (1965). In *Organogenesis*, ch. 6, ed. De Haan, R. L., and Ursprung, H. New York: Holt, Rinehart & Winston.
SWISHER, J. E., and HIBBARD, E. (1967). *J. exp. Zool.*, **165,** 433–440.
SZÉKELY G. (1966). *Adv. Morphogen.*, **5,** 181.
SZENTÁGOTHAI, J. (1961). In *Brain and Behavior*, vol. 1, ed. Brazier, M. A. B. Washington, D.C.: American Institute of Biological Sciences.
VAN DER LOOS, H. (1965). *Bull. Johns Hopkins Hosp.*, **117,** 228–250.
WEILER, I. J. (1966). *Expl Neurol.*, **15,** 377.
WEISS, P. A. (1937). *J. comp. Neurol.*, **66,** 481–535.

For discussion see pp. 67–76.

CELL DIVISION AND MIGRATION IN THE BRAIN AFTER OPTIC NERVE LESIONS

R. M. GAZE and W. E. WATSON

M.R.C. Research Group on the Central Mechanisms of Vision, and Department of Physiology, Edinburgh University

THE ability of severed amphibian optic nerve fibres to regenerate back to their appropriate places in the tectum shows elements of selection of both pathway and destination. In selective regeneration of this sort there are two aspects of the system which have to be considered together—the regenerating fibres and the environment in which they regenerate. The regenerating fibres come either from the original ganglion cells, as in the cut frog optic nerve, or from newly differentiated ganglion cells, as in the adult newt, where the retina degenerates after section of the optic nerve. In this latter situation one should perhaps talk not of "regeneration" of the fibres, but rather of development occurring in an adult animal.

However, the other half of the regenerating system has also to be considered—the environment into which the fibres grow. This environment consists of degenerating axonal fragments, intact neurons and glial elements, and it is this milieu which must provide the structural basis for selective regeneration. We therefore thought it would be of interest to investigate the cellular changes that might take place in the brain during the interval between section and regeneration of an optic nerve. With this object we have followed DNA synthesis, cell division and cell migration in the eye and tectum of the newt, using radioautography with tritiated thymidine.

METHODS

Forty μc of tritiated (^{3}H) thymidine (specific activity 5 c/m-mole) were administered intraperitoneally to normal newts (*Triturus cristatus*) and, by a single injection, in different animals at 1, 2, 3, 4, 6, 7, 14, 21, 28, 33, 35, 42, 56, 63, 70 and 77 days after intraorbital section of the left optic nerve. The operation included avulsion (under ether anaesthesia) of the extraocular muscles and blood vessels and nearly complete removal of the eye from the orbit, followed by its replacement. The animals were killed at intervals of 1, 2, 4, 5, 7, 14, 21, 35, 91, 168 and 175 days after injection. The eyes and brain were removed, fixed in Carnoy, embedded in paraffin and sectioned at 1 or 5 μ; radioautographs were prepared according to the method described by Watson (1965). The present results are based on the study of 105 animals.

Hay and Fischman (1961) showed that after the intraperitoneal injection

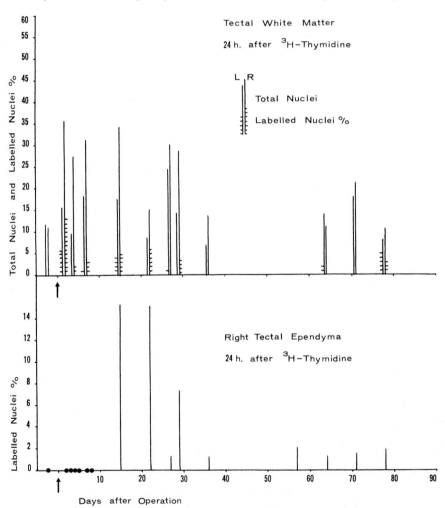

FIG. 1. *Top:* Tectal white matter. Total nuclear count and percentage of labelled nuclei, on right and left side, at various intervals after section of the left optic nerve.
Bottom: The percentage of labelled ependymal cells in the right tectum, 24 hours after administration of [³H]thymidine at various intervals after section of the left optic nerve.
The arrows indicate the times of operation.

of 5 μc of [³H]thymidine in newts, the thymidine disappeared from the plasma within three hours. It is likely, therefore, that in the present experiments the thymidine was initially available for cellular uptake only for a period of hours.

Counts of labelled cells presented in this paper are mean counts from four sections, separated from each other by at least 10 μ. These counts are given as percentages of the total nuclear counts. In the tectal white matter, absolute counts of all nuclei are given as well as percentages of labelled nuclei.

RESULTS

Normal tectum

In the normal newt, killed one day after administration of [³H]thymidine, the tectum showed only very sparse labelled cells in the white matter (less than 1 per cent) and none in the grey matter or ependyma. This absence of labelled cells was also found when the animal was killed 91 days after injection of the isotope. Thus, in unoperated animals, the level of cell division in the tectum is very low.

Tectal white matter after section of the optic nerve

Within 48 hours of section of the left optic nerve (this being the earliest time at which examination was made), the number of cells in the tectal white matter was greater on the right side than on the left (Fig. 1). Some of these cells on the right side were labelled when [³H]thymidine had been given 24 hours before the animal was killed. This cellular increase in the tectal white matter is shown in Fig. 2, which also shows the absence of labelled cells anywhere else in the tectum. The existence among the excess cells of unlabelled cells suggests that the increase in cell numbers had begun before the label was given, 24 hours after operation.

Tectal ependyma after section of the optic nerve

In animals killed 24 hours after the administration of [³H]thymidine, labelled cells were first found in the contralateral tectal ependyma 15 days after operation, when some 15 per cent of the ependymal cells were labelled (Fig. 1). Later than 22 days after operation the number of ependymal cells labelled 24 hours after injection decreased, but they were still detectable 78 days after operation (Fig. 1).

When [³H]thymidine was given to animals between three and seven days after operation, no labelled ependymal cells were found if the animals survived only a further 24 hours (Fig. 1). If, however, they survived until about the tenth postoperative day—between seven and four days after the administration of isotope—some ependymal cells became labelled (Fig. 3). This finding may indicate either that the [³H]thymidine has become

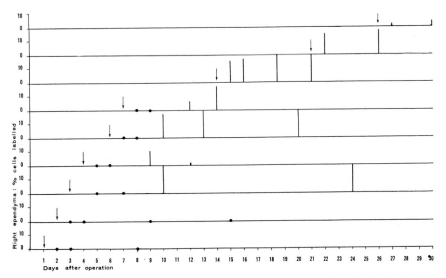

FIG. 3. Composite diagram showing the appearance of labelled ependymal cells in the right tectum after injection of [³H]thymidine at various times after operation. Each horizontal strip represents the appearance of label at various intervals after injection of [³H]thymidine. The times of administration are indicated by the arrows. Filled circles represent no labelling. Vertical bars show the extent of labelling. For compactness, counts of labelled cells above 20 per cent have been treated as if they were 20 per cent.

FIG. 2. Radioautograph of tectum two days after operation and 24 hours after administration of [³H]thymidine. The right tectum is on the right. There is increased cellularity of the right tectal white matter compared to the left and some of the excess cells are labelled. There are no labelled cells in the ependyma or grey matter (compare FIG. 7). Some cells in the meninges and choroid plexus are labelled. In all figures the bar represents 100 μ unless otherwise stated.

FIG. 4. Radioautograph of right tectal grey matter ten days after operation and seven days after administration of [³H]thymidine. Occasional cells in the ependyma (lower surface in figure) are labelled.

FIG. 5. Radioautograph of right tectum ten days after operation and four days after administration of [³H]thymidine. Many of the ependymal cells are labelled, as are some cells in the white matter.

FIG. 6. Radioautograph of tectum 15 days after operation and 24 hours after administration of [³H]thymidine. The right tectal ependyma is becoming labelled.

FIG. 7. Radioautograph of tectum 15 days after operation and 24 hours after administration of [³H]thymidine. The right tectum is on the left. Many ependymal cells and cells in the adjacent grey matter are labelled. The white matter on the same side also shows more cells than on the left.

FIG. 8. Radioautograph of a different section from the same animal as FIG. 7. The spread of the labelled cells to the deep grey matter adjacent to the ependyma is shown.

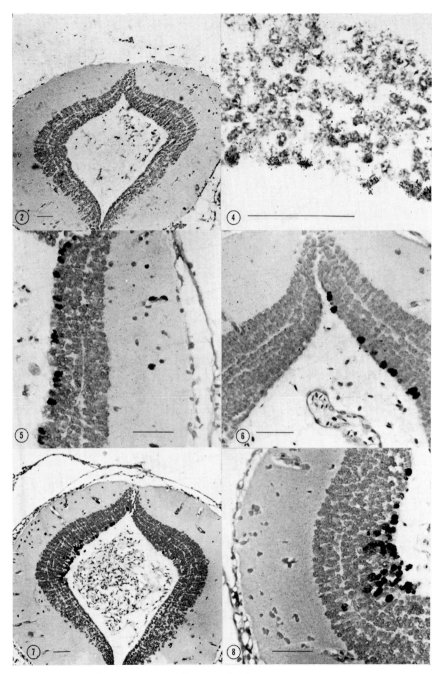

FIGS. 2 and 4–8

incorporated shortly after administration into acid-soluble oligonucleotides in the cells which show delayed labelling, or that the [³H]thymidine has become incorporated by these cells about the tenth day, from isotope released by degenerating labelled cells. No ependymal labelling was found when [³H]thymidine was given within two days of nerve injury, even when followed up to the 15th day after operation.

The earliest ependymal labelling was demonstrated nine days after operation, when the [³H]thymidine had been given at four days. Fig. 4 shows early ependymal labelling occurring seven days after injection and ten days after operation. Another example of early ependymal labelling is shown in Fig. 5, where the [³H]thymidine was given six days after operation and the animal was killed four days later.

Immediate (24-hour) ependymal labelling was seen in these experiments from 15 days onwards (Figs. 3 and 6). When more than 24 hours was allowed to elapse after the administration of [³H]thymidine at 14 days after operation, the tectal label was not confined to the ependyma but included also the immediately adjacent cells in the periventricular grey matter. Some animals, from which radioautographs were prepared 24 hours after the isotope had been given on the 14th day after operation, also showed label extending into the periventricular grey matter. Fig. 7 shows such a case, where the injection was made 14 days after operation and the animal was killed 24 hours later. Fig. 8 illustrates another section from the same brain, where the labelled cells spread several cells deep beyond the ependyma. At this stage after operation, the uptake of [³H]thymidine into DNA is associated with numerous mitoses in the same cell layers (Fig. 9).

With longer intervals between the administration of [³H]thymidine and radioautography, the cellular labelling extends until it may be uniformly

---

FIG. 9. Radioautograph of part of the right tectal ependyma from the same animal as FIGS. 7 and 8, showing mitosis in one of the ependymal cells. The bar represents 15 μ.

FIG. 10. Radioautograph of right tectum 20 days after operation and 14 days after administration of [³H]thymidine. There is an almost continuous distribution of labelled cells in the ependyma and subjacent grey matter, but the outer grey matter is not labelled.

FIG. 11. Radioautograph of right tectum 105 days after operation and 91 days after administration of [³H]thymidine. The ependymal and adjacent label is still present but the rest of the grey matter and the white matter are not labelled.

FIG. 12. Ependymal cells with their processes. Heidenhain's iron haematoxylin.

FIG. 13. Radioautograph of normal retina 24 hours after administration of [³H]thymidine. Two adjacent cells are labelled at the ciliary margin of the retina. No other cells in the section are labelled. The choroid is heavily pigmented. The bar represents 50 μ.

FIG. 15. Radioautograph of normal retina 91 days after administration of [³H]thymidine. The label covers all three retinal layers for some distance from the ciliary margin. See FIG. 14b.

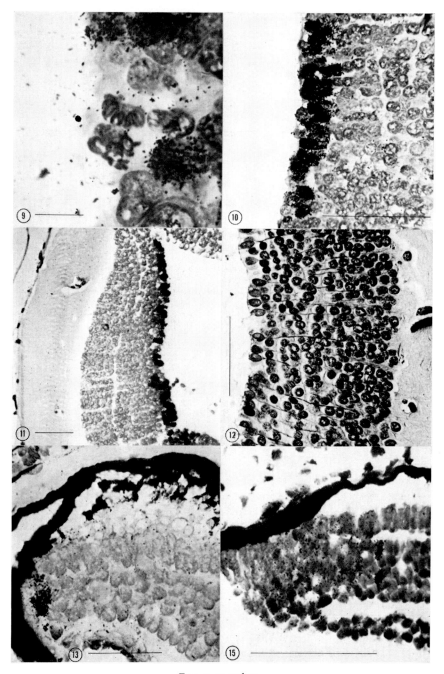

Figs. 9–13 and 15

several cells deep all round the ventricular margin of the tectum (Fig. 10). However, as shown in Fig. 10 (animal injected six days after operation, killed 14 days later), the labelled cells do not appear to move further out into the grey matter. They remain concentrated in the ependyma and periventricular grey matter and are still evident there 91 days after administration of the isotope. This is well shown in Fig. 11, where the [^{3}H]thymidine was given 14 days after operation and the animal was killed 91 days later. The labelling of the ependyma and deep grey matter is obvious, although in this animal all traces of the increased cellularity of the tectal white matter had disappeared.

The highest count of labelled ependymal cells, 24 hours after the administration of [^{3}H]thymidine, occurred 15 days after operation and was 15 per cent. Ependymal labelling increased with time after injection and the highest count observed was 51 per cent (in a single section). However, the true figures would be considerably higher than this, because in each count a variable proportion of unlabelled, non-tectal ependyma was included. The tectal ependyma merges imperceptibly with the ependyma of the ventral midbrain. However, direct observation showed that, at intervals of more than several days from injection, labelling in the tectal ependyma approached 100 per cent.

The ependymal cells in the newt consist of a bulbous, nucleated expansion at the ventricular border and a thick stem passing outwards through the grey matter (Fig. 12). The thick stem of the cell gives off mossy filaments in the grey matter and also in the white, where the filaments may terminate only at the outer surface of the brain (Herrick, 1942). These ependymoglial cells thus provide an extensively ramifying framework for the other tectal elements.

Ependymal cells in the contralateral wall of the diencephalon also became labelled after section of the optic nerve. The timing was similar to that in the tectal ependyma. Sections in the rostral part of the diencephalon, near the chiasma, showed groups of labelled ependymal cells situated ventrally in the wall of the ventricle; sections in the caudal part of the diencephalon, near the tectum, showed groups of labelled ependymal cells situated more dorsally in the ventricular wall.

In the present experiments the observations were not continued long enough after operation to show any changes that might be associated with ingrowth of the regenerating optic nerve fibres. Retinae examined up to 70 days after operation showed various degrees of disorganization (30 animals); all retinae examined 95 or more days after operation (14 animals) showed restitution of normal retinal structure. Probably, therefore, regeneration of optic nerve fibres had not occurred before 70 days after operation.

Normal retina

The retina in the adult newt continues to grow by adding cells to all three cell layers at the ciliary margin. Radioautography 24 hours after the administration of [³H]thymidine gives sections which in many cases show

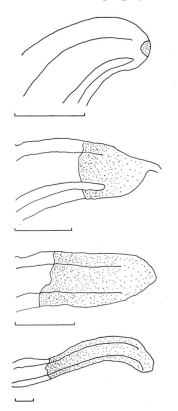

FIG. 14. Distribution of label in retinae of normal and operated eyes at various intervals after administration of [³H]thymidine.

(*a*) Normal eye 24 hours after injection; two adjacent cells are labelled at the ciliary margin. This is the same retina as shown in FIG. 13. The section has been inverted in the diagram.

(*b*) Normal eye 91 days after injection; the labelling has spread to all three cell layers of the retina. The label is most concentrated furthest away from the ciliary margin. This retina is shown in FIG. 15. The section has been inverted in the diagram.

(*c*) Normal eye 168 days after injection; the labelled cells have spread even further in all retinal layers and again the label is most concentrated furthest away from the ciliary margin. This retina is shown in FIG. 16. The section has been inverted in the diagram.

(*d*) Operated eye, 231 days after operation and 175 days after injection. The label here has spread a considerable distance along the retina and is most concentrated furthest away from the ciliary margin. This retina is shown in FIGS. 21 and 22.

Note the difference in scale between *a, b, c* and *d*.

only one or two labelled cells at the retinal margin (Figs. 13 and 14). This indicates that the retina is terminated at the ciliary margin by a more or less continuous ring of cells synthesizing DNA. The retina grows by division of these cells. Retinal growth may be slow, as is shown by the finding that in some animals only a single cell was labelled at the retinal margin seven days after injection. However, 91 days after administration of [³H]thymidine, labelled cells have spread some distance along the retina, involving all three cell layers (Figs. 14 and 15). When studied even later after injection (Figs. 14 and 16), the labelled region of the retina is seen to have spread

3*

further from the ciliary margin. In Figs. 14, 15 and 16 it may be seen that the most intensely labelled cells are those furthest from the ciliary margin; thus the region of active cell division is the ciliary margin and the cells furthest away divide less and maintain their nuclear label in less diluted form.

Retina after section of the optic nerve

After section of the optic nerve the retina degenerates, and this process was usually complete within 21 days under the conditions of these experiments. The earliest signs of retinal degeneration were seen at the fundus, while the retina close to the ciliary margin maintained its integrity for longer periods. When degenerative changes were visible at the fundus it was commonly found that these were initially restricted to the ganglion cell and bipolar cell layers and that the receptors were more resistant to disintegration.

Retinae from animals injected with [^{3}H]thymidine four or seven days after operation and radioautographed one to seven days later, showed labelled cells at the ciliary margin in some cases. Animals injected four to 42 days after operation showed, 91 days after injection, complete labelling of the entire re-formed retina.

Because of the considerable variability in the time-course of retinal degeneration and regeneration from one animal to the next, it was not possible to obtain an entirely consistent picture by radioautography. However the probable course of events is as follows: initially degeneration is most marked in the fundus, which may be reduced to a single layer of cells

---

FIG. 16. Radioautograph of normal retina 168 days after administration of [^{3}H]thymidine. The label extends further than in FIG. 15 and is most concentrated furthest from the ciliary margin, which is at the left of the photograph. See FIG. 14c.

FIG. 17. Radioautograph of retina 57 days after operation and 24 hours after administration of [^{3}H]thymidine. The fundus consists of a single layer of cells, many of them labelled. At the ciliary margin there is a region of active proliferation.

FIG. 18. Radioautograph at higher magnification of the same section shown in FIG. 17, showing the labelled region at the ciliary margin.

FIG. 19. Radioautograph of retina 72 days after operation and two days after administration of [^{3}H]thymidine. Both fundus and ciliary margin are labelled. In the fundus the outer retinal layer is unlabelled while the inner layer is undifferentiated and labelled.

FIG. 20. Higher magnification of part of the section shown in FIG. 19, showing the labelled region at the ciliary margin.

FIG. 22. Radioautograph of retina 231 days after operation and 175 days after administration of [^{3}H]thymidine. This retina is shown in FIGS. 14d and 21. The concentration of label is greatest furthest away from the ciliary margin, which is to the right in the photograph. The edge of the unlabelled part of retina is at the left of the photograph.

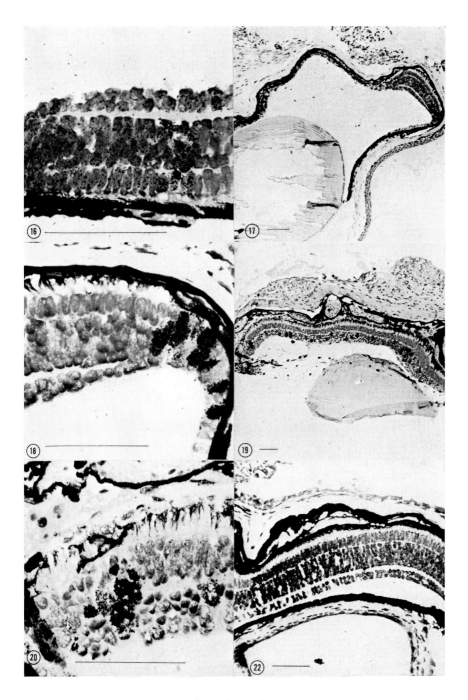

Figs. 16–20 and 22

while the cells at the ciliary margin survive. At this stage, pigment cells at the fundus invade the remains of the retina and from about 20 days after operation the entire fundus shows much labelling within 24 hours of injection. Fig. 17 shows such a situation. In this animal, [^{3}H]thymidine was administered 56 days after operation and the animal was killed 24 hours later. The fundus is represented by a monolayer of cells, many of them labelled. The ciliary margin shows a region of more or less normal retina which, judging by the rapidity of labelling of cells there (Fig. 18), has probably all regenerated from the active cell region at the edge of the retina.

Another animal, radioautographed after a longer interval, showed a different picture. Here the active region of regeneration at the ciliary margin was also seen, but the fundus showed two distinct layers; the receptor layer appeared to be intact, well differentiated and unlabelled, while the bipolar and ganglion cell layer showed multiple labelling and had not, as yet, differentiated into its two constituent layers (Fig. 19). This animal was injected with [^{3}H]thymidine 70 days after operation and radioautographed two days later. The region of active cell division at the ciliary margin is shown in Fig. 20, which also shows the contrast between the well-formed, unlabelled outer layer and the disorganized, much-labelled inner retinal layer.

Although these results indicate that, early in regeneration, the fundus itself is much labelled and presumably the site of much cell division, the most strikingly consistent finding in all animals was the extent of cellular activity at the ciliary margin of the retina. Late after section of the optic nerve, this region tended to show extensive labelling occurring within 24 hours after the isotope was given. And it seems likely that the greater part of the fully regenerated retina is made up of cells which came from this region of the eye. In an animal injected 56 days after section of the optic nerve and radioautographed 175 days later, the well-formed retina showed labelled regions extending out from the ciliary margin and nearly meeting in the fundus (Figs. 14 and 21). The distribution of this label (Figs. 14, 21 and 22) showed that all these labelled cells had come from the marginal region of the retina. A similar extent and distribution of retinal label was seen in the two other newts, injected respectively at 56 and 63 days and radioautographed at 231 days after operation.

DISCUSSION

The adult newt retina is continually growing. Stone (1959) pointed out the existence at the ciliary margin of the retina of a group of "undifferentiated" cells which he thought could probably start dividing again if called upon to do so. The present results show that these cells are continuously, if slowly, dividing, and that over a period of, for example, 100 days, they may add significantly to the extent of the retina.

This raises the problem of where the fibres from this newly synthesized retina go. The anatomical basis of the retino-tectal projection has been laid down long previously, in embryonic life. Unless the tectal end of the fibre projection is to get progressively more cramped, the newly growing optic nerve fibres must go to newly developing areas of the tectum. It is not known at present whether the tectum also continues expanding, to take these fibres.

In the initial stages of retinal degeneration, administered [^{3}H]thymidine appears only at the ciliary margins, while the unlabelled fundus shows progressive degenerative changes. When animals were injected four days

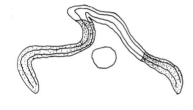

FIG. 21. Diagram of section of retina 231 days after operation and 175 days after administration of [^{3}H]thymidine. This retina is shown in FIGS. 14d and 22. The concentration of label on each side of the retina increases towards the edge of the labelled area furthest from the ciliary margin. The bar represents 1 mm.

after operation and radioautographed 91 days later, the entire retina was seen to be fairly uniformly labelled. This uniformly distributed retinal labelling could thus have come from the retinal margins, where active proliferation was in progress at the time of injection.

From approximately 20 days after operation, labelled cells may be found scattered throughout the fundus as well as at the ciliary margins. The eventual fate of these labelled fundal cells is not obvious, since the three radioautographs made 91 days or more after injection and where the animals were injected, respectively, at 56, 56 and 63 days after operation, showed the greater part of the re-formed retina to have come from the ciliary margins (Fig. 21). It is possible that, as in the developing frog retina (Glücksmann, 1965), the central retinal cells degenerate, to be replaced by those from the marginal region.

The observation that much of the regenerated retina has its origin at the ciliary margin suggests an explanation for the results of Cronly-Dillon (1967), who found that, some 150 days after removal of the retina in newts, the retino-tectal projection from the central area of the retina was almost normal while that from peripheral retina was not. The present experiments indicate that the reason for this is likely to be that the central retina has been formed first by movement of cells in from the margins, while the peripheral retina is still being formed. In the regenerated eye, central retina is "older" than peripheral retina.

The present evidence concerning the origin of the regenerated retina may also help to explain some of the rather perplexing findings of Grafstein and Burgen (1964). These authors found various peculiarities in the newt retino-tectal projection four to seven months after retinal operations which included removal of the retina with or without excision of the fundal pigment epithelium and with or without rotation of the peripheral retina and pigment epithelium. The assumption of Grafstein and Burgen that central retina derives from central pigment epithelium is not, in view of the present results, justified. Furthermore, the consistency of rotation of the central and peripheral parts of the visual projection, which they reported, is to be expected if the entire retina originates from the ciliary margin. Finally, these authors allowed four to seven months to elapse between operation and recording, in order to be sure that the retino-tectal connexions had become well established. The present results suggest that this time is unlikely to be long enough for this purpose.

The origin of most of the regenerating retina from the ciliary margin calls attention to the problem of the origin of ganglion cell specificity. The spatial differentiation of the ganglion cells must presumably either inhere in the dividing cells, or be transmitted to the cells from, probably, the pigment epithelium. Since ganglion cells in the fundal region may originate peripherally, and pass across the retina to their eventual destinations, it would be interesting to know more about their specificity characteristics at different stages in their careers.

The renewal of the ependymoglial cells that occurs in the contralateral tectum after section of the optic nerve differs from the gliosis that is seen in the optic nerve, optic tract and tectal white matter. The increased cellularity in the optic pathway occurs within two days of nerve section. Ependymoglial cells start dividing about nine or ten days after operation and, over a period of several days, virtually all of these cells become labelled after the administration of [^{3}H]thymidine. Ependymoglial cells are the main structural elements in the newt tectum (Herrick, 1942), and this labelling pattern indicates that, after section of the optic nerve in these animals, a considerable change occurs in the glial structure of the tectum. The tectum into which the regenerating fibres will grow is not the same in cellular constitution as it was before section of the optic nerve. It will be interesting to find out whether this ependymoglial reaction occurs also in other animals possessing similar structures (e.g. frog: Lázár and Székely, 1967) and whether it has a major effect on the mechanism of fibre regeneration.

SUMMARY

The cellular response in the CNS to section of the optic nerve in the newt (*Triturus cristatus*) was studied by radioautography with tritiated thymidine. In normal animals and in animals at various intervals after nerve section, 40 μc of tritiated thymidine (specific activity 5 c/m-mole) was administered intraperitoneally. The animals were killed at different times after injection, and serial sections of the eyes and brain were radioautographed.

Unoperated control animals showed only sparse and sporadic labelling of tectal cells, even when followed for up to 91 days after administration of the isotope. Normal retinae showed sections in which only a single labelled cell could be seen at the ciliary margin 24 hours after injection; by 168 days after injection, a block of cells, covering all three retinal cell layers, had become labelled.

Gliosis was seen in the optic tract within two days of the optic nerve being sectioned. From ten days after operation, profuse labelling of the contralateral tectal ependyma occurred. The retina, which degenerated after nerve section, regenerated initially over a period of 95 days. Later distribution of label showed that most of the regenerated retina had come from the periphery.

REFERENCES

CRONLY-DILLON, J. R. (1967). *J. Physiol., Lond.*, **189**, 88–89P.
GLÜCKSMANN, A. (1965). *Archs Biol., Liège*, **76**, 419–437.
GRAFSTEIN, B., and BURGEN, A. S. V. (1964). In *Topics in Basic Neurology*, pp. 126–138, ed. Bargmann, W., and Schadé, J. P. [*Progress in Brain Research*, vol. 6.] Amsterdam: Elsevier.
HAY, E. D., and FISCHMAN, D. A. (1961). *Devl Biol.*, **3**, 26–59.
HERRICK, C. J. (1942). *J. comp. Neurol.*, **77**, 191–353.
LÁZÁR, G., and SZÉKELY, G. (1967). *J. Hirnforsch.*, **9**, 329–344.
STONE, L. S. (1959). *J. exp. Zool.*, **142**, 285–308.
WATSON, W. E. (1965). *J. Physiol., Lond.*, **180**, 741–753.

DISCUSSION

Szentágothai: Can animals re-learn under these circumstances?

Sperry: We see almost no indication of any re-learning in the sense of a correction of the reversed vision. Our frogs went for months and Stone's salamanders for years without any indication of re-learning.

Szentágothai: In our experience *Triturus vulgaris* never re-learns the use of a reversed visual field; it does not even seem to become aware of anything being wrong. Conversely *Triturus cristatus* soon "recognizes" the error and begins to make clumsy efforts for correction. The difference between the two species appears to be in the time of developmental determination of the extraocular muscles. The transplanted eyes in *Triturus vulgaris* were completely immobile— the operations having been made in early larval stages (25–28)—whereas those of *Triturus cristatus* regained complete movement. It was assumed that the latter animal received information about its reversed visual field through conflicting proprioceptive and optic feedback resulting from eye movement. When there is no eye movement, as in *Triturus vulgaris*, the most important cue is lacking (Szentágothai, J., and Székely, G. [1956]. *Acta physiol. hung.*, **10**, 44–55).

Sperry: Whenever there is a good eye and a reversed eye, the system of the good eye will predominate.

Szentágothai: Our procedure in both cases was entirely similar: the lure was always presented to the eye with reversed visual field and re-learning was judged from the performance in this situation.

Eccles: How do the fish manage to survive when they are always missing their food?

Sperry: One can place the food on the nose or even in the mouth. The salamanders can feed in the dark if the meat is thrown on the bottom and frogs and toads can be force-fed.

Eccles: The point about feedback is important. At all levels of animal behaviour the feedback must be a very significant guide to action, as it were. The fact that the good eye leads to success and the bad eye fails must eventually give patterns of behaviour, but this is nothing more than a conditioned behaviour.

Sperry: It would be impossible to get any learning without feedback. Learning to use the good eye and ignore the bad eye doesn't involve very much detailed re-learning. It is half-way between maze learning and learning upside-down vision.

Székely: Several years ago in an experiment I removed the temporal half of the eye primordium and replaced it by a nasal half, in salamander embryos (Székely, G. [1954]. *Acta physiol. hung.*, **6**, Suppl. 18). The eye therefore developed from two nasal halves. From similar experiments performed on toads (Gaze, R. M., Jacobson, M. and Székely, G. [1963]. *J. Physiol., Lond.*, **165**, 484–499), we know that optic fibres from both the original and grafted nasal poles occupy the posterior part of the tectum, which is the normal place of termination of nasal fibres; and the anterior part of the tectum, which is the place of termination of temporal fibres, was invaded by optic fibres from the middle part of such eyes. The salamanders with operated eyes produced reversed optic reflexes from the visual field of the grafted half of the retina. However, after a few erroneous responses the animals refused to react from either half of the visual field. Probably by the aid of a visual feedback, they learned that they were performing erroneous reflexes, and in the absence of a mechanism for correcting the errors they could not but switch off the reflexes of the operated eye.

Hamburger: In connexion with the reactivation of mitosis in the ependymal layer of the tectum, Dr. Gaze, did you imply that at first a degeneration occurs under the influence of the degenerating optic fibres and then a replacement?

Gaze: Ten days or so after the operation intense labelling and mitosis of the ependymal cells suddenly start. Mitosis is not preceded by degeneration as far as I know.

Hamburger: Do you imply, then, that normally the optic fibres inhibit mitotic activity in the ependymal layer? If so, do they do so directly? Do any fibres reach that far down?

Gaze: No fibres reach that far down. I don't know whether the optic fibres normally inhibit ependymal mitosis. To find this out we would have to have two optic nerves going to one tectum and cut one, or something like that. At present we have no alternative explanation for the activation of mitosis.

Young: Some of the ependymal cells reach the surface, so couldn't they detect the degenerating fibres?

Gaze: Some of the processes do go all the way up to the surface and they come in contact with the degenerating fibres immediately. But the ependymoglial cells do not show this nuclear reaction when the degenerating fibres hit them. That is, within 24 hours of optic nerve section one finds the gliosis in the white matter, but the ependymoglial nuclear changes only appear ten days later. We have no explanation for this latency. It may be that it takes this much time for information to pass back down the ependymoglial processes to the nuclei.

Kollros: If the eye is removed in the frog larva the mitotic activity of the tectal ependyma decreases considerably (Terry, R. J., and Gordon, J., Jr. [1960]. *J. exp. Zool.*, **143**, 245–257), and after temporary interruption of the optic nerve over ten or 15 days, once the connexion is restored, the ependyma increases its activity so as to catch up with the control side (McMurray, V. M. [1954]. *J. exp. Zool.*, **125**, 247–263). The relationship of the innervation here is that optic nerve fibres, if anything, enhance during normal development the activity of mitotic cells in the tectal ependyma.

Hughes: In the 1940s, Professor Sperry, your explanation of the maladaptive responses was that the cutaneous local sign is imparted onto sensory fibres by a process analogous to myotypic specification. In your recent writings you regard nerve fibres growing towards an organ as having their "numbers" already on them. Have you altered your ideas about the cutaneous local sign?

Sperry: Certainly the cutaneous fibres are already labelled in part as far as the cranial nuclei, etc., are concerned. The suggestion was that it is just the local terminations with a dermatome that are specified by the skin. In the dorsal-ventral reversal of trunk skin, we can't really be sure that ventral fibres don't grow dorsally. But it doesn't look as though leg fibres will grow so far up as the thoracic region to reach limb buds transplanted there. It would therefore appear that cutaneous integumental induction, or specification of the cutaneous fibres for local sign, does occur in the frog. It wouldn't make a great deal of difference so far as behaviour mechanisms go whether the fibres are all specified autonomously and grow out and find their proper spot, or whether they grow out at random and then become specified by cutaneous contacts. It is just a different way of working out the same developmental problem, but it all comes out in terms of very refined neuronal specificity with individual markers.

Hughes: There seems to be a fundamental difference between the two cases: in one the fibres are indifferent until they reach their peripheral terminations, while in the other they are already determined before they get there.

Sperry: That may be critical from the embryologist's standpoint, but from the standpoint of function it doesn't make much difference which way it is done. I wouldn't be surprised to see it done both ways in different species or in different systems in the same organism. I suspect that the cutaneous fibres are partially specified for, say, the cranial or trigeminal or thoracic areas and so on, and maybe for dermatomes or for ventral and dorsal rami, but within a particular cutaneous area it is unlikely, though not impossible, that the individual fibres should have to grow out and find a particular spot in the skin.

Hughes: So you envisage the determination happening either before or after the fibre gets to its place of termination?

Sperry: Yes, or a combination of the two—except that I don't see any way of interpreting the limb transplants otherwise than by postulating that the fibres have limb specificity induced as a result of contact with the skin.

Eccles: It might help if we realize that the movements Professor Sperry showed in his beautiful film are the overall effect of a whole nervous system, with control and feedback and so on, but 100 per cent connexions are not needed to get these results. They may look 100 per cent, but a great deal of neuronal machinery is present to ensure that some organized action happens and only a majority vote of information is needed to get this to happen.

Székely: In some of Professor Sperry's experiments there seems to be a contradiction between the results and the interpretation. After the cross-union of the mandibular and ophthalmic nerves, the animal behaves as if the trigeminal ganglion cells had been specified prefunctionally and the innervated periphery is unable to re-specify them. Is this true?

Sperry: It is true in all animals and in all species. If you wait until everything is specified and fixed, then conditions are similar to those in man.

Székely: When the belly skin of a frog is transplanted to the back, the animal behaves as if the spinal ganglion cells had obtained their specificity from the innervated skin. Why is it that the cranial sensory ganglion cells cannot be specified, or modulated, and spinal ganglion cells can be modulated by the innervated periphery?

Sperry: Cranial nerves and cranial ganglion are much more highly specified. If you cross-transplant the mandibular and the ophthalmic nerves at early enough stages the same re-specification effect is seen. The experiment you asked about was one where an ophthalmic nerve was transplanted into the mandible and

DISCUSSION

mandibular responses were seen. If this is done early enough apparently this is what happens, although there have been so few experiments I wouldn't swear by it.

Székely: In one experiment I grafted an eye into the lower jaw of young *Xenopus* larvae. From the grafted eyes, which were evidently innervated by the mandibular nerve, corneal reflexes like those of the normal eye could be evoked. In this case the cranial sensory ganglion cells were modulated by the innervated periphery.

Sperry: They are modulated if the experiment is done *early* enough. Most of the crosses you are referring to were on eft stage salamanders (newts). If one goes back far enough, the modulation in the face area occurs, just as it does in the limbs. This is consistent with what you found with the cornea. It is just a question at what stage you get modulation and at what stage the fibres are fixed. If one goes back far enough in the embryo I am sure a stage will be found where everything can be "modulated".

Székely: At which stage did you do the mandibular-ophthalmic nerve cross-union experiment?

Sperry: I don't remember any more. It must be mentioned in those same papers.

Gaze: From everything we have recorded it appears that under no circumstances does 100 per cent regeneration occur. We have never recorded a retino-tectal projection in *Rana temporaria* which is 100 per cent normal. In most cases it is pretty grossly abnormal and in all cases it is slightly so. This is not to say, of course, that after a lengthy period of regeneration, one may not get a fair approximation to a normal retino-tectal map.

Eccles: I was referring more to the other cutaneous experiments of Professor Sperry in this respect, though I think my suggestion would also apply to the various visual organ responses. The visual responses look so clean and clear and yet one doesn't have to postulate that 100 per cent regeneration occurs.

Sperry: I tried to use *Rana pipiens* for optic nerve studies and had to give up—the optic nerve of our adult Ranidae did not regenerate very well. In the Hylidae one gets much more rapid and complete regeneration.

Gaze: One can certainly get excellent restitution of behaviour. Our problem with these frogs is that we may sometimes get more than 100 per cent regeneration, not less: the ipsilateral fibres may go back to the wrong side as well as the normal fibres.

Sperry: We have had cases where in the chiasma some of the fibres fail to go across but they terminated apparently at the right spot in the ipsilateral tectum.

Gaze: Yes, but one has a supernormal, not a normal situation, after that. The animal can still function after a fashion and the specificity of regrowth has been well and truly manifested, but it is not a normal situation in many cases; an excess of fibres may be coming from the other eye, for instance.

Sperry: How long after metamorphosis were your experiments carried out?

Gaze: These would be two- to three-year-old frogs.

Sperry: One might get much better regeneration immediately after metamorphosis when the frogs are younger. We simply gave up on *Rana pipiens* because it took months, as you report, whereas the tree frogs and other species were functioning again in a month. With young goldfish it was only two weeks for recovery of vision.

Szentágothai: When a limb is transplanted into the vagal skin region and becomes innervated, then touching the limb doesn't give a corneal reflex. However, when the hand is amputated, a regeneration blastema soon develops which on stimulation gives the corneal reflex. This reflex disappears as soon as the leg has fully regenerated. If the hand is amputated once more, the same process occurs again. According to your suggestion, this would imply that the connexions giving rise to the corneal reflex had to be grown out and then withdrawn according to whether the periphery is a blastema or a real leg.

Sperry: How long does it take a blastema to grow?

Székely: In the salamander it takes two days to grow and it persists for two to three weeks. The corneal reflex appears within two days. The grafted limb is innervated by the vagus sensory ganglion. The distance between the vagus centre and the corneal reflex nucleus is approximately 2 to 3 mm. in the brain.

Sperry: Could not the fibres become "despecified" and readjust their synapses in two days? The cut surface, as we all know, tends to be a pretty sensitive area on the skin and corneal reflexes are threshold phenomena; possibly the blastema has a very tender tip that is extra sensitive.

Singer: The salamander's blastema of regeneration is very sensitive. Already within two days after amputation transected nerve fibres have grown into the wound area; they invade the regenerating epithelium in great numbers and even penetrate to the free surface (Singer, M. [1948]. *J. exp. Zool.*, **103**, 189–210). These epithelial fibres are sensory ones and are thus exposed to mechanical stimulation, rendering the blastema a very sensitive structure. Indeed, the animal responds immediately and rather violently to stimulation of the regenerate with a needle prick.

Eccles: I think we are coming round to a reasonable physiological explanation of these observations. There is always this barrage going into the nervous system and one has only to move the operation of the neural circuitry slightly one way or the other to get one type of reaction or another.

Székely: I agree with Professor Singer that a regeneration blastema, as well as a cornea, is very rich in nerves. Moreover, from the investigations of E. D. Hay (1960. *Expl Cell Res.*, **19**, 299–317) one can conclude that both tissues have the same type of sensory innervation. These observations call, indeed, for a physiological explanation rather than for neuronal specificity in the interpretation of the corneal reflex experiments.

Young: That won't work when it comes to a detailed pattern of function. You are talking about corneal reflexes, which are pretty unspecific responses, but the eye muscle responses described by Professor Sperry are all highly specific. It all depends on the degree of precision of the function concerned.

Eccles: Even these highly specific reflexes that look so clean in their operation may not need a 100 per cent neuronal mechanism to be like that. There are many feedback patterns of inhibition and so on that result in a clean action, although the neuronal machinery is very complex.

Hughes: What happens when a newt eye is grafted into the trunk and stimulated?

Kollros: When it is grafted in place of the forelimb, there is no corneal reflex in the salamander larva (Kollros, unpublished data).

Hughes: Does it give any kind of sensory response?

Kollros: It is sensitive. The animal will move as a result of a touch on the grafted eye. I did not test to see whether it was more sensitive than the adjacent skin, although from Professor Singer's remarks this might well be so. There is no indication that it has the specificity of corneal skin.

Walton: Does the order of latency suggest that the cutaneous adaptive reflexes you describe are monosynaptic or polysynaptic, Professor Sperry? What happens if you divide the spinal cord above the segment?

Sperry: I would have guessed the responses were polysynaptic. We decerebrated some of the animals but I don't believe we ever transected the cord.

Székely: I repeated the skin reflex experiments on some 30 or so frogs, rotating the skin all the way around on some of them, so that the whole back was covered with belly skin and the whole belly with back skin. The misdirected wiping reflexes did not work so nicely in these frogs as in those in which the trunk skin was rotated only on one side. In most cases it was difficult to find a point from which misdirected reflexes could be evoked. Most of the rotated skin area gave rise to normal reflexes. If any kind of specific regrowth or specific modulation of the nerves had occurred, the frogs in which the whole skin was rotated should have shown much better misdirected reflexes than those in which only half of the belly and back skin was rotated.

Sperry: This would probably depend on the stage at which you did the experiment, and also perhaps on how many layers of the dermis are included.

Székely: I followed exactly the same procedure as N. Miner did (1956. *J. comp. Neurol.*, **105**, 161–170), in the tadpole. I used *Rana esculenta* while she used *Rana pipiens*, but they look pretty much the same.

Buller: Was it only after you had repeated Miner's experiments and obtained the same answer that you did a total replacement and didn't get the same answer?

Székely: Yes, that is correct.

Szentágothai: Professor Sperry, would your observation of optic fibres growing into the tectum rule out the notion that a few pioneering fibres growing out to the opposite tectum, but otherwise at random, are accepted if they reach the appropriate place and rejected if they don't? In this case one could imagine that later outgrowing neighbours of the pioneering fibres grow along those that had been accepted and not along those that had been rejected. Correct connexions could be established by such a mechanism. In other words, do the nerve fibres have some prior information on exactly which place to go to, or is the site of termination established by this trial and error procedure?

Sperry: I would expect a lot of this exploratory process to go on locally in the plexiform layer, but it would be pretty hard to account for the choice of the main pathways to get to the tectum. Over large areas of the tectum I wouldn't expect a fibre that went into the anterior ventral quadrant to be able to correct itself and get to a proper zone in the posterior ventral quadrant. I think the evidence shows a good deal of selection in the choice of pathways that lead to the target. After it gets to its general area, there is probably quite a bit of local searching about. There is certainly a tremendous growth of fibres in the plexiform layer in the regenerated area.

Eccles: Is it possible that there are residual polarized chemical changes so that the fibres are regenerating not into a uniform environment, but into one where something has been left behind that guides them?

Sperry: The assumption is that the whole pathway has its chemical labels on it that serve as signs in a sense that direct the fibres "this way" or "that way".

Eccles: The glial cells may be helping in this case. Professor Szentágothai made the point that only a few fibres need to find the way and the others can follow, each one providing information, once a successful target-finding cell gives the information to the others.

Sperry: I am not sure what you gain by that. What do you have in mind?

Szentágothai: The rough direction of outgrowth might be determined, but otherwise target-finding might be based on a trial and error procedure that could work also with random outgrowth.

Sperry: If they grow at random the fibres that go wrong would have to degenerate all the way back through the brachia, because the lateral and medial tracts are selectively filled. The lateral tract is filled from the ventral half of the retina but not from the other half, and the medial tract is filled or avoided depending on which half of the retina is regenerating. Also they are segregated back at the optic chiasma. I assume that the tissues are marked all along the way. If they are crossed at the point where the brachia divide they will regain their original channels but further dorsally the lateral fibres don't have much chance to test and sample the dorso-ventral differences.

Eccles: I thought the idea was that there was a minimum number of up or down labels.

Sperry: All one needs logically are two gradients—and I assume there is at least a third, radial gradient as well. The mammalian visual system is very precise and with just these three gradients one can account in principle for the whole organization of the mammalian system, which, you see, poses some new problems since in mammals the nasal half-field of the retina is projected in register with the temporal half-field of the other eye.

Szentágothai: Let us assume that only 1 per cent of the total fibre population consists of pioneering fibres and that the others would only follow the courses given by those pioneering fibres that have made appropriate contacts. The few pioneering fibres that have not been able to establish appropriate contacts and therefore were not followed by the later-coming masses of fibres would not show up histologically.

Sperry: But how do the other fibres know? What is the difference between ten fibres that have made a successful contact and the whole system of glial cells?

Szentágothai: There is ample histological evidence (Rajkovits, K. [1953]. *Acta morph. hung.*, **3**, 43–49) that regenerating fibres which have established appropriate contacts at their terminals start growing quickly, whereas those which have failed do not. One could imagine that something in connexion with this growth or maturation stimulates its neighbours in the central stump to grow along. The neighbours would then have a fair, or certainly a better, chance of reaching the appropriate target.

Sperry: You are in fact saying that the other 90 per cent don't go by chance but by selective growth?

Szentágothai: Yes. I do not deny that finally the optic fibres find their appropriate sites of ending, but I would be interested in the mechanism by which this is achieved.

Prestige: In the regenerating neuromuscular system, J. T. Aitken, M. Sharman and J. Z. Young (1947. *J. Anat.*, **81**, 1–22) and others have shown that the nerve fibres branch extensively above the level of section, and that they grow down and remain branched until they reconnect with the muscle: only then are the branches lost. Could this be a basis for selective regrowth?

Sperry: If the nerve is allowed to regenerate and is fixed and examined at intervals before vision is restored, then one finds that before the nerves ever reach their terminals they fill up selectively one or the other tract.

Drachman: Dr. Gaze, you have shown that the radioactive label first appears in the ependymal nuclei, in the optic tectum. Later, labelled nuclei are found further out. Are any of these labelled cells known to be neuronal?

Gaze: Many of them are obviously not, although I can't prove that any of them are definitely neural. I don't as yet know how to differentiate neurons from glial elements effectively in newts. They are probably ependymoglial cells which are no longer ependymal. At the moment I can't prove this, because I haven't been able to see the large thick stems reaching out from them towards the surface. These ependymoglial cells are dividing, and some of the mitoses are labelled. How does a cell, which has such processes going up to the surface, round up for mitosis? We know nuclear mitosis must occur and we can actually see it, but presumably there is cell division also taking place. Conceivably the extra nucleus is extruded in a bud from the main ependymoglial process, but we have no information on this.

Hughes: We should not assume that over the whole field of nerve and end-organ relationships the same explanation applies, or the same proportion of explanations. The regeneration of the optic nerve probably stands at an extreme of fineness of relationship, but I am sure that this doesn't apply to, say, nerve-muscle contacts in a limb. The primary development of these contacts, as far as I can see, follows exactly what Professor Szentágothai has said about the pioneer fibres and the others which enter later; a great deal of selection occurs finally. We shall not advance the subject by assuming what isn't necessarily true, that the method of regeneration of nerve fibres into the optic tectum applies to everything else.

Eccles: I think Professor Buller and I would agree that when nerves regenerate to the muscles there is very little specificity between one muscle and another. In fact nerves to slow muscles will regenerate quite well when crossed to fast muscles, and *vice versa*, though perhaps the connexions are not 100 per cent.

Levi-Montalcini: Both you and Dr. Gaze showed that regenerating nerve fibres select the appropriate pathway in the optic tract long before they reach their matching nerve cells in the tectum, Professor Sperry. Should this be taken as evidence that some biochemical agent diffuses into the medium and guides the regenerating nerve fibres towards the matching cells in the optic tectum?

Sperry: If the optic nerve is examined at daily intervals after section one finds selective outgrowth from the beginning—not a full growth followed by selective degeneration (Attardi and Sperry, 1960, 1963, *loc. cit.*).

Székely: Can you show the termination of the optic fibres?

Gaze: Normally speaking, the fibres, as you showed, select their appropriate pathway in the tract, whereas when the fibres are deliberately directed into the wrong pathway, they still find their correct destinations.

Sperry: Once they get into that peculiar parallel layer they appear to be stuck, and they are forced to follow that pathway for reasons which we don't yet understand. Dr. Hibbard and I are still working on this problem.

DEVELOPMENT OF LIMB MOVEMENTS: EMBRYOLOGICAL, PHYSIOLOGICAL AND MODEL STUDIES

G. Székely

Department of Anatomy, University Medical School, Pécs

SINCE the pioneer experiments of Detwiler and his co-workers (Detwiler, 1936), it has been known that an extra limb innervated by the thoracic segments of the spinal cord is unable to perform coordinated movements in salamanders. If the limb is grafted close enough to, and receives innervation from, the brachial plexus, it moves in a coordinated manner and in synchrony with the normal forelimb. This finding led Weiss (1941) to propose the theory of neuronal modulation as a basis of coordinated limb activity. To summarize briefly, Weiss assumes that each muscle has some constitutional specificity by which it is distinguished from all other muscles. Motor nerves invade the limb at random, and the muscles reached by chance impart their specificity to the motor neurons via the innervating axons. The motor neurons, in turn, select their central synapses in accordance with their acquired specificity. As a final result, motor neurons with identical specificity deliver impulses simultaneously.

A number of observations made on limb movements under various experimental conditions (Székely, 1966) have raised some doubt about the validity of Weiss's suggestion. This doubt has been further reinforced by a preliminary effort to record muscle activities, both from normal and from synchronously moving grafted limbs, in freely moving animals. While some of the homologous limb muscles (such as the extensor ulnae) were active simultaneously in both limbs, other homologous muscles (for example, brachialis, extensor digitorum communis) contracted with a considerable phase shift relative to each other. Since the phase shift was in the range of 100 to 300 msec., it was too small to be detected by watching the animal, and the movements of the normal and grafted limb appeared to be synchronized.

These contradictory observations have led to an attempt to approach the problem of the development of limb movements from a different direction.

NERVOUS ACTIVITIES CONTROLLING THE RHYTHM AND MOVEMENT PATTERN OF THE LIMB ARE INHERENT IN THE LIMB SEGMENTS OF THE CORD

It has been known for many years that deafferentation of the limbs does not seriously affect the general pattern of walking in amphibians. From the observation that a completely deafferented toad maintained the basic pattern of motor coordination, Weiss (1936) concluded that limb movements were centrally controlled. His conclusion met with opposition from Gray (1950), who could not confirm Weiss's observation and found that intact motor and sensory nerves of at least one spinal segment were essential for coordinated limb movements. He concluded that ambulatory rhythm was dependent on rhythmic excitation from the periphery.

To extend the study of whether or not the ambulatory rhythm is dependent on rhythmic peripheral excitation, we have made use of Weiss's (1950) "deplantation" technique to isolate various segments of the spinal cord. A tunnel was made into the dorsal fin of the salamander larva. A piece of spinal cord was introduced into it and the mouth of the tunnel was closed by a limb graft. This limb became innervated by the transplanted spinal cord. If the cord graft was taken from the brachial region, the limb displayed walking-type coordinated movements either spontaneously or in response to light touches of the area around the cord graft. But if the spinal cord was of thoracic origin, only irregular twitches of the limb muscles were seen. With mechanical stimuli these twitches could be brought into tetanic muscle contractions, but never into the kind of coordinated movements mentioned above. Subsequent histological examination suggested that preserved structural organization of the grafted spinal cord was necessary for coordinated movements to occur.

The experiment shows that spinal cord segments from the limb level are capable of controlling coordinated limb movements even if they are isolated from the rest of the nervous system and receive non-rhythmic stimuli.

Other experiments with these segments of the cord indicate that the fore- or hindlimb pattern in the act of walking is also determined by the centre.

Thoracic segments of young salamander larvae were replaced by brachial or lumbosacral segments, and a pair of extra limbs was grafted into the same region (Székely, 1963). Extra limbs innervated by the grafted brachial segments moved in phase with the normal forelimbs, and those innervated by lumbosacral segments moved in phase with the hindlimbs. The origin of the extra limbs—whether fore- or hindlimbs—was irrelevant with respect to the rhythm of movements; however some minute differences could be observed in the character of the movements. The knee moves very little in a normal hindlimb, and similarly, when a forelimb was innervated by lumbosacral segments, during its movements in synchrony with the normal hindlimbs, the elbow remained relatively motionless. This effect was more pronounced with the limb movements of chickens. Straznicky (1963) replaced the brachial segments by lumbosacral segments in chick embryos. After hatching the wings moved in parallel with the legs on the same side, but were unable to perform wing-like fluttering movements. Therefore, both the rhythm and the character of limb movements are determined by the spinal cord, and the innervated periphery has no detectable effect on the function of the centre.

OUTGROWING MOTOR FIBRES REACH THEIR PROPER DESTINATION BY A PREFERENTIAL SELECTION OF PERIPHERAL PATHWAYS

Sperry and Arora (1965) and Mark (1965) have recently reported experimental findings which indicated selectivity in the reconnexion of regenerating oculomotor nerve and pectoral fin motor nerve fibres in cichlid fish. Our recent finding is also suggestive of a preferential selection of peripheral pathways by the outgrowing limb motor nerve fibres.

It is known that thoracic spinal cord segments, if transferred in early embryonic stages, are able to replace the function of the brachial segments (Detwiler, 1923). This experiment was repeated using increasingly older embryos for operation (Straznicky and Székely, 1967). If the brachial segments were replaced by thoracic segments before the early tail-bud stage, normal limb movements developed. If the cord was transplanted in increasingly older embryos, movements failed to develop first in the shoulder, then in the elbow, and finally the whole limb remained motionless (Table I). In other words, functional adaptation in the grafted thoracic segments failed to occur in a proximo-distal direction with respect to the anatomical positions of the limb muscles. It is generally believed that the proximo-distal order of the muscles in the limb is represented in a craniocaudal order of the innervating motor neurons in the cord. In view of this belief it was thought that, in these experiments, histological differentiation of the grafted cord was gradually failing to take place—that is fewer and fewer motor neurons were differentiating in the cranio-caudal direction, in parallel with the increasing poverty of limb activity in the proximo-distal direction. To check this possibility, fibres in the two large ventral roots (3 and 4) of the brachial plexus were counted. The data of Table I indicate that the number of fibres decreased in parallel with the poverty of limb activity, and this decrease occurred approximately in the same measure in both roots. If coordinated limb movements require the proper number

Table I

NUMBER OF FIBRES IN THE VENTRAL ROOTS OF GRAFTED THORACIC SEGMENTS

Stage of operation	Number of animals	Fibre counts in ventral roots		Average numbers		Motility		
		Root 3	Root 4	Root 3	Root 4	Shoulder	Elbow	Wrist
Normal embryos	1	415	470					
	2	468	496	435	431			
	3	471	351	**866**		+	+	+
	4	385	406					
23	1	354	281			±	+	+
	2	539	503	408	387	+	+	+
	3	419	405	**795**		±	+	+
	4	320	358			−	+	+
24	1	433	329			±	+	+
	2	292	318	343	343	−	±	+
	3	304	382	**686**		±	±	+
25	1	257	202			±	±	+
	2	231	168	221	171	−	±	+
	3	240	157	**392**		−	−	+
	4	274	164			−	±	+
	5	129	162			−	−	±
26	2	119	115			−	−	±
	3	69	81	105	77	−	−	−
	4	124	—*	**182**		−	−	+
	6	123	188			−	±	+
	7	92	—*			−	−	±

The first rows indicate the number of ventral root fibres in four normal animals; the following rows show the data obtained from animals operated at successively older stages. The bold-face figures are the sum of the average numbers of fibres in roots 3 and 4. The last column indicates the motility of the forelimbs: +, normal; ±, defective movements; −, lack of movement in the corresponding joints.

*Only root 3 was present.

of motor fibres to the corresponding muscle groups, then in those cases, for instance, where only hand movements are seen, the muscles of the shoulder and arm should receive many fewer nerve fibres than the muscles of the forearm and hand. If this is so, it is difficult to understand why most of the outgrowing nerve fibres pass by the proximal muscles and terminate in the distal muscles, unless one assumes that the nerve fibres preferentially select the pathways which lead them to their proper destination.

MOST LIMB MUSCLES ARE REPRESENTED IN EACH OF THE THREE BRACHIAL SEGMENTS

The possibility that most limb muscles are represented in each of the three brachial segments was suggested by the previous experiment. To study the arrangement of motor neurons supplying individual muscles, the brachial spinal cord was electrically stimulated at the points of a rectangular three-dimensional grid (Fig. 1) in lightly anaesthetized axolotls (Székely and Czéh, 1967). Capillary electrodes with a tip diameter of 3-5 μ, filled with 0.65 per cent sodium chloride solution, were used for stimulation. Square voltage impulses of 1-50 v amplitude and 1 msec. duration were passed through the electrode with a frequency of 1/sec. The evoked muscle contractions were observed under a dissecting microscope. With this kind of stimulation, single muscle contractions, or contractions of well-defined parts of a muscle, could be elicited. From the results of several measurements (comparing the depth of the electrode with different thresholds; marking the places of lowest thresholds; distance measurements on histological sections), it was concluded that a muscle could be activated if the electrode tip was within a distance of 50 μ from the body or the main dendrites of the innervating motor neuron(s). The results of stimulations are summarized in Fig. 1, which shows the points from which contractions of individual muscles could be evoked with the lowest amplitude of stimulus.

Despite the many short-comings of the method and presentation of the results, which will not be discussed here, inspection of Fig. 1 indicates several things: (*i*) The topographical relations of the muscles are only inaccurately reflected in their central representation. (*ii*) Both the synergist and antagonist muscle groups of the three principal joints of the limb are almost evenly represented throughout the brachial cord. (*iii*) The only apparent order is that the muscles involved in protraction and elevation of the shoulder and in flexion of the elbow are represented cranially to the corresponding antagonist (retractor-depressor-extensor) muscle group, but the two fields overlap extensively. The forearm muscles are represented in a smaller cranial and in a larger caudal field. Extensor and flexor muscles are equally represented in both fields. (*iv*) An interesting feature, not apparent in the diagram, is the grouped arrangement of motor neurons innervating individual muscles. This is suggested by the following observations. The representation of, for instance, the extensor ulnae covers

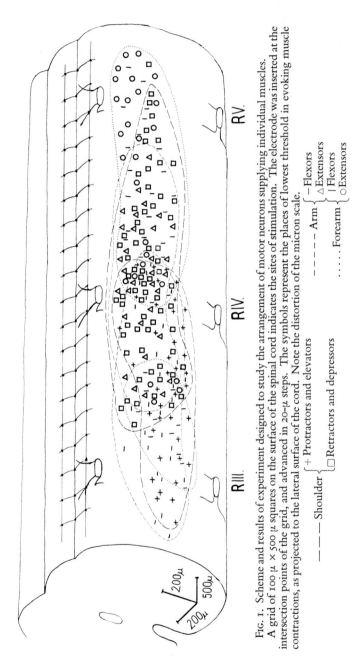

FIG. 1. Scheme and results of experiment designed to study the arrangement of motor neurons supplying individual muscles. A grid of 100 μ × 500 μ squares on the surface of the spinal cord indicates the sites of stimulation. The electrode was inserted at the intersection points of the grid, and advanced in 20-μ steps. The symbols represent the places of lowest threshold in evoking muscle contractions, as projected to the lateral surface of the cord. Note the distortion of the micron scale.

---- Shoulder { + Protractors and elevators / □ Retractors and depressors
---- Arm { - Flexors / △ Extensors
.... Forearm { | Flexors / ○ Extensors

an area of about 6 mm. in length. In the course of stimulation, as we proceeded caudally from root III, activity in this muscle first appeared with the electrode 1,500 μ behind the root. A number of other muscles, both synergist and antagonist, could be activated from the same level. At the next level the extensor ulnae was found again, mostly together with the same muscles as before, but perhaps also with one or two new muscles. On stimulation 500 μ further caudally, the extensor ulnae could not be activated. It reappeared again at the next level, but with a different set of muscles. The muscle disappeared and reappeared again for shorter or longer distances, and these small areas of representation of the extensor ulnae were shared with a number of other muscles. This holds true for the other muscles as well. Since the diagram is a compound representation of the results obtained from 15 animals, the grouped representation of individual muscles is indistinct in Fig. 1, because the different levels of stimulation do not indicate identical points in the different spinal cords. The stimulation experiments give the impression that motor neurons supplying a number of different muscles are arranged in small groups alternating with other groups of motor neurons innervating different sets of muscles.

The problem of this apparently random arrangement of motor neurons can probably be approached more successfully from a functional point of view. A myochronogram was constructed for this purpose (Fig. 2), illustrating the contraction sequence of different muscle groups in one cycle of walking, derived from slow-motion pictures of the salamander. It is clear that muscles concerned in protraction of the limb (shoulder protractors and elevators, elbow flexors), are acting nearly simultaneously. They are represented in a common field. The same is true for the muscles which retract the limb (shoulder retractors and depressors, elbow extensors), and these are represented in another field. The two fields overlap extensively. The forearm muscles show a double phase of activity and, as could be concluded from experiments recording muscle action potentials, the muscle activity during one of these phases is weaker.

It is of interest to compare this function with their double field of representation. Further analysis in this direction is hampered by the difficulty of determining unambiguously the function of the muscles. The list of muscles in Fig. 2 indicates their possible cooperation in limb movements by deriving their functions from their anatomical positions. The list reveals that most limb muscles participate in protraction, as well as in retraction of the limb. Some of them (such as the latissimus dorsi and coracobrachialis) may act either as synergists or as antagonists in various phases of walking. Their classification as protractors or retractors, therefore, does not necessarily mean that they perform only the designated action. A preliminary effort to record muscle potentials in freely moving animals has shown that the active periods of such antagonistic pairs of muscles as the brachialis and extensor ulnae, or the dorsalis scapulae and pectoralis, overlap extensively during ambulation. The synergist-antagon-

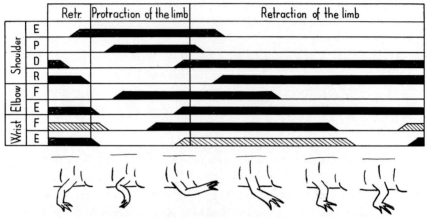

ELEVATION: **Dors. scapulae, Scapulohumeralis,** Latiss. dorsi
PROTRACTION: **Acromialis, Pectoralis** (ant.), Scapulohumeralis, Flex. compositus
DEPRESSION: **Rectoralis** (med.), **Coracobrachialis,** Flex.compositus
RETRACTION: **Coracobrachialis, Pectoralis** (post.), **Latiss, dorsi,** Dors.scapulae
FLEXION: **Brachialis, Flex.comp.,** Flex. carpi rad., Ext.carpi uln.
EXTENSION: **Ext. ulnae**
FLEXION: **Flex. carpi rad., Flex. dig. comm., Flex. carpi uln.**
EXTENSION: **Ext. carpi rad., Ext. dig. comm., Ext. carpi uln.**

FIG. 2. Myochronogram of the right forelimb of a salamander in the act of walking over solid ground, derived from slow-motion pictures.
The hatched stripes are for a period of weak activity of wrist flexors and extensors, taken over from a preliminary experiment recording muscle action potentials from the limbs of freely moving animals. The series of drawings under the diagram show the positions of the right forelimb in subsequent phases of a walking cycle. The list gives the movements of three joints, in the same order as the initial letters in the diagram. Muscles in ordinary type only partially contribute to the movements indicated.

ist relationship of the muscles continuously changes in different phases of a walking step. In a single attempt to stimulate the spinal cord at the points of a much closer grid (100 × 100 μ squares), where no muscles can theoretically escape detection, the combination of muscles represented in the small areas mentioned earlier appeared to be organized in such a manner that muscles predominantly present in these areas may act as either synergists or antagonists. It seems, therefore, that a curious kind of "functional map" is superimposed on the "anatomical map" of individual muscles in the spinal cord. As the previous experiments suggest, this structure is capable of controlling the highly complicated limb movements without any afferent information from the limb.

While the scarcity of the experimental data does not allow much insight into the intricate structure of the spinal cord segments concerned with limb movement, it may show the way to further studies on this problem. The suggestion that the outgrowing nerve fibres select their proper route in the limb, and the finding that almost every limb muscle is represented in every little piece of the brachial cord, unambiguously explains the experimental finding that an extra limb innervated by any branch of the brachial plexus moves in synchrony with the normal limb. Actually, this

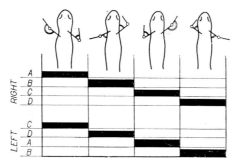

FIG. 3. Simplification of a walking step to four consecutive contraction phases of elbow extensors (A), shoulder retractors (B), elbow flexors (C) and shoulder protractors (D). (By courtesy of the Publishing House of the Hungarian Academy of Sciences.)

is exactly what one would expect, without assuming any kind of neuronal modulation. The experiments presented indicate a highly complicated neural structure in the spinal cord, capable of a delicate integrative function in the control of limb movements. The study of such complex structures calls for new methods.

ATTEMPTS TO SIMULATE AN ASSUMED NERVOUS MECHANISM CONTROLLING COORDINATED LIMB MOVEMENTS, WITH THE AID OF MODEL EXPERIMENTS

Technical model

It is obvious from the foregoing experiments that the limb-moving spinal cord segments are capable of delivering rhythmic outputs to non-rhythmic inputs. A simple logical network can be composed to simulate this function (Székely, 1965). Let us simplify the walking movement to four consecutive contraction phases of the elbow extensors, shoulder retractors, elbow flexors and shoulder protractors, as shown in Fig. 3. Each group of muscles is supplied by a separate pool of motor neurons

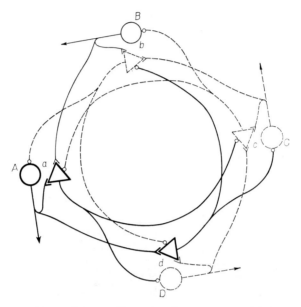

Fig. 4. Diagram illustrating the connexions of four motor neurons (circles) and four inhibitory neurons (triangles). Forked endings (—<) denote excitatory synapses, circle endings (—○) inhibitory synapses.

The principle of the network's function is the following. If A fires, it excites neurons a and d. These inhibit neurons D, C, c and b. The inhibition of b allows for repetitive discharges of neuron A, and the suppression of c removes the inhibition from B. Being thus in a responsive state, B will be activated by the continuous stream of impulses in the next moment, and its discharges will inhibit A and D through inhibitory neurons b and a, respectively. The inhibition of d by b puts C in a responsive state and this latter will discharge in turn; and so on. (By courtesy of the Publishing House of the Hungarian Academy of Sciences.)

(circles), and to each motor neuron an inhibitory neuron is assigned (triangles). The neurons are interconnected as indicated in Fig. 4. If the motor neurons receive a continuous stream of impulses, the network delivers a rhythmic output in the sequence *A-B-C-D*. Neurohistological and neurophysiological data are in favour of the assumed neuronal interconnexions and interactions.

The network was submitted to tests using artificial neurons constructed by Jenik (1962). The *A-d*, *D-c*, *C-b* and *B-a* connexions turned out to be wrong. After they had been disconnected and a regular or random series of impulses applied to the motor neurons, the network delivered an output pattern very much like the simplified myochronogram in Fig. 3. A considerable overlap was experienced in the activity of the neighbouring motor neurons, and the extent of overlap depended on the amplitude and frequency of the input pulses.

From these experiments it is obvious that the function simulated is not

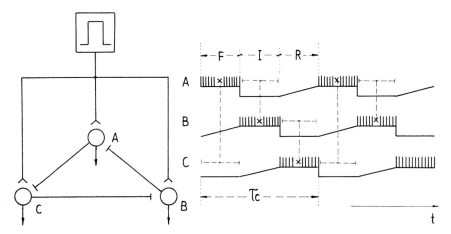

Fig. 5. Network composed of three artificial neurons (circles).
The neurons are driven from a common source of stimulation (forked endings); the interconnexions (terminating in bars) exert inhibition. Arrows indicate the output channels. On the output diagram, F is the firing period, I the inhibited period and R the recovery period. The latter indicates the time which elapses between the cessation of inhibition and the beginning of activity, and acts as an adjustable timing device inserted between the elements. Broken lines starting from the F periods show the inhibitory action on that neuron to which it is connected. τ_c indicates the length of one cycle, which begins with the firing of neuron A and ends with the inhibition of neuron C. t indicates the time axis of the diagram.

that of the limb. However, this finding initiated a study into the nature of networks containing recurrent cyclic inhibition (Kling and Székely, 1967), of which only a short review is given here to show the versatile functional properties of such networks.

The construction of the artificial neurons allowed their use as output (motor) neurons and inhibitory neurons at the same time. This made possible the construction of simple networks of such "compound" elements. Fig. 5 shows a network of three elements. Each element is driven from a common source of stimuli and the cyclic interconnexions exert inhibitory effects. On the output diagram both depolarizing (upward) and hyperpolarizing (downward) membrane-potential changes are shown. The function of the network can easily be read from the diagram. If *A* fires, it inhibits *C* and removes the inhibition from *B*. Then *B* recovers from inhibition, starts firing and inhibits *A*, thus removing inhibition from *C*, and so on. It is obvious that the pattern of the output (the length of the firing periods) depends on the length of the recovery period. With increasing or decreasing input parameters (amplitude and/or frequency) the length of the recovery period can be made shorter or longer respectively. This means that an element would start firing after a shorter or longer recovery period and display a longer or shorter firing period, but the length of one cycle (τ_c) would not change.

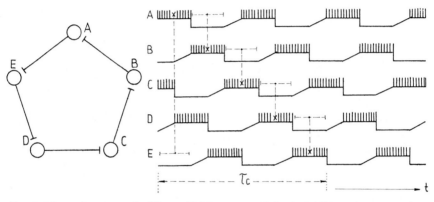

FIG. 6. Network composed of five artificial neurons. Only the inhibitory interconnexions are shown.

Fig. 6 shows a network of five elements. In the output pattern the firing period is twice as long as the recovery period, and the next cycle starts before the previous one is over. Because of the interaction of elements, the *F* and *R* relation cannot change freely.

In Fig. 7 another network of five elements is shown in which two inhibitory connexions arise from each element. This network delivers the output pattern shown in the diagram: *R* is twice as long as *F*. If the two inhibitory connexions are directed differently (Fig. 8), the output pattern remains the same but the sequence of discharges changes. A similar network with three inhibitory connexions from each element will obtain an output pattern with equal *F* and *R* periods.

From these few examples it is obvious that in addition to the length of the recovery periods, the number of elements and of the inhibitory connexions contribute to the establishment of the output pattern. From these relationships simple equations can be derived, and the output patterns of networks composed of any number of elements can be computed as a function of the length of the recovery periods, provided that the elements are ideally uniform (every R is equal), and the network is composed in a

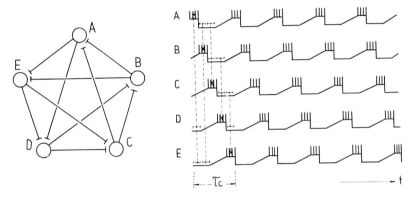

FIG. 7. Network composed of five artificial neurons in which two inhibitory connexions arise from each element. These are connected to the first and to the second neighbour in an anticlockwise direction.

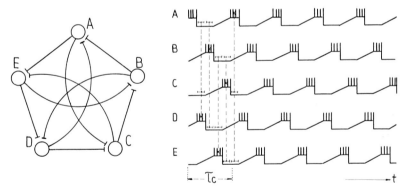

FIG. 8. The same network as in Fig. 7, but the inhibitory connexions are directed to the first and the third neighbour in an anticlockwise direction. Note that the discharge sequence of the elements is changed: A—D—B—E—C.

symmetrical manner (each element gives rise to equal numbers of inhibitory connexions).

It is not difficult to see that by introducing unequal R values and asymmetric structures, so that the numbers of inhibitory connexions are different, countless variations in the output pattern can be produced. An example is shown in Fig. 9 for an asymmetric structure, in which one additional inhibitory connexion is introduced in a simple network of five elements.

The output pattern and the sequence of discharges is dramatically changed (cf. Fig. 6).

To get closer to reality, we also studied networks in which separate output (excitatory) and inhibitory elements were employed. Fig. 10 shows such a network of five pairs of elements. One can readily imagine the innumerable further variations of the output pattern which can be achieved by varying the *R* values and the structure.

The duration of the recovery period acts as a timing device inserted between the elements. Since its length depends on the input parameters, both excitatory and inhibitory, it may be taken to represent the host of events which happen before a hyperpolarized nerve cell starts firing. In other words, the recovery period is intended to simulate the complicated

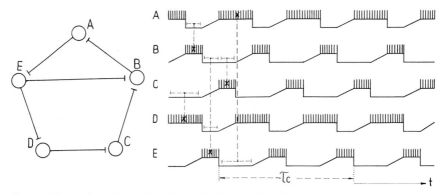

FIG. 9. Network similar to that shown in Fig. 6. An additional inhibitory connexion is introduced from *E* to *B*. Note the changes in the firing pattern and sequence.

integrative function of the nerve cell. The inhibitory elements are of the Renshaw type of nerve cells. In model experiments we have assumed that these were driven exclusively by the motor neurons. It is well known that Renshaw neurons in the spinal cord receive both excitatory and inhibitory stimuli from primary afferent fibres, from other spinal interneurons and from a number of supraspinal centres. These input pulses can very effectively contribute to the establishment of the output pattern by controlling the assumed recovery period of the inhibitory neurons.

As already mentioned (p. 81), motor neurons innervating individual muscles seem to be arranged in small groups in which both synergist and antagonist muscles are represented according to a functional scheme. If one replaces these groups by pools of neurons—consisting of motor neurons, and excitatory and inhibitory interneurons—on which afferent and descending impulses finally impinge, and connects them in the above manner, this would offer a relatively simple structure capable of controlling the versatile activity of a limb merely by quantitative variations of the input.

Biological model

As mentioned earlier, thoracic cord segments are unable to move a limb. The only difference that one can see between thoracic and limb

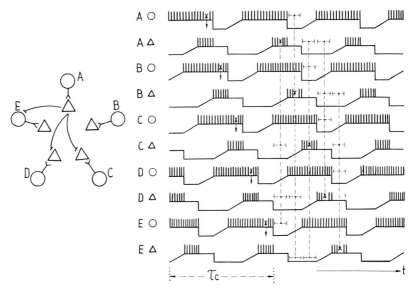

FIG. 10. Network composed of separate output (circles) and inhibitory (triangles) neurons.
Only the output neurons receive stimuli from a common source. These, when they are active, drive the inhibitory neurons (note the connexions with forked endings in the network, and arrows in the output diagram). Each inhibitory neuron gives rise to three inhibitory connexions (terminating in bars), of which only one set is shown in the figure. Their inhibitory actions are indicated by the broken lines on the output diagram.

segments under the microscope is that the number of neurons is much larger in the ventral horns of the limb-moving segments. In the experiment replacing the brachial cord by thoracic segments, the number of neurons decreased in parallel with decreasing limb activity. By simulating the rhythmic activity of thalamus on an electronic computer with networks containing recurrent inhibition, Andersen and Rudjard (1964) found that the larger the number of neurons being inhibited after the discharge of one cell, the more pronounced was the rhythmic activity. These observations suggest that the number and the closeness of neurons in a given area might play a decisive role in generating rhythmic activity. We have therefore attempted to increase the number of neurons in the thoracic segments (Székely and Straznicky, 1967).

The experiments were performed on salamander neurulae. A piece of the medullary plate was cut out from the region of the prospective thoracic cord and transplanted under the medullary plate in the same region of another neurula. Two medullary plates were, therefore, layered on each other. A pair of extra limbs were transplanted at the level of the medullary plate operation at an early larval age. After about two weeks from the onset of normal limb movements, rhythmic movements appeared in the grafted limbs in about half the operated animals. These movements were far from coordinated limb movements; nevertheless, clear biphasic movements—presumably indicative of rhythmic discharges from the cord—could definitely be seen, sometimes in every joint, or more often in one or two joints of the grafted limb. Histological studies are in progress to see whether the extent and rhythmicity of the movements can be correlated with the number of motor neurons.

SUMMARY

From the finding that an extra limb innervated by brachial spinal cord segments moves in synchrony with the normal forelimb, Weiss proposed the theory of neuronal modulation as a basis of coordinated limb activity. Some recent findings are inconsistent with this theory. On the basis of the following experiments on urodeles an alternative interpretation of the neural mechanisms underlying coordinated limb activity is suggested:

(1) Spinal cord transplantation reveals that the nervous activities determining the rhythm and movement pattern of the limb are inherent in the limb segments of the spinal cord.

(2) Thoracic cord segments can replace the brachial segments in every respect, if they are transferred in early embryonic life. When the thoracic segments are taken from increasingly older embryos, movements fail to develop first in the shoulder, then in the elbow and finally in the wrist. Parallel with this functional deficiency the number of ventral root fibres decreases evenly throughout the grafted segments. The findings suggest that the outgrowing motor fibres reach their proper destination by preferential selection of the peripheral pathways.

(3) Stimulation of the brachial segments with microelectrodes reveals that practically every limb muscle is represented in each of the three brachial

segments. This finding and the selectivity in outgrowth of motor fibres accounts for the complete synchronous movement of an extra limb innervated by any branch of the brachial plexus.

(4) Model experiments with artificial neurons show that networks with recurrent cyclic inhibition generate rhythmic output in response to continuous regular or random input. The function of such networks is exceedingly versatile and suggests an easy way to stimulate the assumed nervous mechanism controlling coordinated limb movements.

(5) In experiments increasing the neuroblast material of presumptive thoracic spinal cord segments in salamander neurulae, extra limbs grafted into the same region display rhythmic movements.

REFERENCES

ANDERSEN, P., and RUDJARD, T. (1964). *Nature, Lond.*, **204**, 289–290.
DETWILER, S. R. (1923). *J. exp. Zool.*, **37**, 339–393.
DETWILER, S. R. (1936). *Neuroembryology: An Experimental Study*. New York: Macmillan Co.
GRAY, J. (1950). *Symp. Soc. exp. Biol.*, **4**, 112–126.
JENIK, F. (1962). *Ergebn. Biol.*, **25**, 206–245.
KLING, U., and SZÉKELY, G. (1967). In preparation.
MARK, R. F. (1965). *Expl Neurol.*, **12**, 292–302.
SPERRY, R. W. and ARORA, H. L. (1965). *J. Embryol. exp. Morph.*, **14**, 307–317.
STRAZNICKY, K. (1963). *Acta biol. hung.*, **14**, 145–155.
STRAZNICKY, K., and SZÉKELY, G. (1967). *Acta biol. hung.*, **18**, 449
SZÉKELY, G. (1963). *J. Embryol. exp. Morph.*, **11**, 431–444.
SZÉKELY, G. (1965). *Acta physiol., hung.*, **27**, 285–289.
SZÉKELY, G. (1966). *Adv. Morphogen.*, **5**, 181–219.
SZÉKELY, G., and CZÉH, G. (1967). *Acta physiol., hung.*, **32**, 3–18.
SZÉKELY, G., and STRAZNICKY, K. (1967). In preparation.
WEISS, P. (1936). *Am. J. Physiol.*, **115**, 461–475.
WEISS, P. (1941). *Comp. Psychol. Monogr.*, **17**, 1–96.
WEISS, P. (1950). *J. exp. Zool.*, **113**, 397–461.

DISCUSSION

Hughes: J. S. Nicholas and D. H. Barron in 1935 (*J. comp. Neurol.*, **61**, 413–431) stimulated spinal nerves III, IV and V of *Ambystoma* and afterwards stimulated the ventral roots separately. On stimulation of III, the reaction was mainly arm abduction and V gave digit flexion. On the other hand stimulation of the ventral roots gave very little indication of selective distribution: there was movement in all segments of the limb from each root. Nicholas and Barron did not conclude much from this, but it suggested that it was central coordination that was all-important in limb movement. That seems to fit in beautifully with what you have been describing, Dr. Székely. In the urodeles, the internal pattern of coordination is probably more important than in other tetrapods.

Kerkut: At stage 26, when just wrist movement is seen, did you try cutting off that limb, getting regeneration and then seeing what that limb would do, Dr. Székely? If you could study the effect of controlled stimulation, it would be nice to see if you got movement of the whole limb.

Székely: I did not do this, because the animals started dying at the time of metamorphosis.

Gutmann: Is the evidence for selective outgrowth sufficient? Is the number of nerve fibres in the nerves innervating different muscles reduced in the same way or is the reduction more pronounced in some of the nerves? The reduction of nerve fibres might be related to a decrease of selective outgrowth (or specificity) but it could also be related to other factors.

Székely: In a few preliminary investigations in collaboration with Dr. K. Straznicky, we determined the number of motor fibres which left the nerve trunks for the shoulder muscles, for the arm muscles and for the forearm muscles. From this we found a relation between the number of fibres and extent of movements in the corresponding joints. In animals with only wrist movement we found a reduced number of fibres going to the shoulder and arm muscles, and a nearly normal number of fibres going to the forearm muscles. This made me conclude that the outgrowing fibres had some cue to find their proper pathways. These investigations are still going on.

Hník: Did you measure the diameter of the regenerated fibres? The phase shift of about 200 msec. between normal and grafted limbs might be due to a slower conduction rate. Peripheral nerve regeneration in very young animals is exceptionally poor (Romanes, G. J. [1946]. *J. Anat.*, **80**, 117). Probably not only is the number of nerve fibres reduced, but also their diameter is very much smaller—50 per cent or even less of the normal size (see Zelená, J., and Hník, P. [1963]. *Physiologia bohemoslov.*, **12**, 277).

Székely: I did not measure fibre diameters. However, I do not think that a slow conduction velocity in consequence of a reduced fibre diameter would account for the 200 msec. phase shift between normal and grafted limb muscles. Other muscles, for example extensor ulnae, contracted exactly in phase. The difference we found between the two kinds of muscles was that the representation of the extensor ulnae extended all over the brachial segments of the cord. Therefore, if the supernumerary limb was innervated, say by root IV, this muscle had a good chance of receiving a proper innervation; while those which contracted with a phase-shift relative to normal muscles might have been innervated by "foreign" nerve cells.

Hník: Did you say that in the forearm the agonist and antagonist muscles contracted simultaneously?

Székely: No. I said they were active during both retraction and protraction, but with some phase shift and with a considerable overlap in their activity.

Piatt: Did you analyse the numbers of primary and secondary motor neurons and correlate this with good, bad or indifferent limb movement? In an old piece of work (1949. *J. exp. Zool.*, **111**, 1–25) I found that poor limb movement was usually correlated with a great deficiency or absence of secondary motor neurons in the brachial segments of the cord. In some work I am doing now on reconstitution of the brachial segments of the spinal cord in *Ambystoma* both the secondary motor system and limb movement are very poor.

Székely: I did not investigate this question.

Eccles: What do you mean by primary and secondary motor neurons?

Piatt: This was something which was brought out primarily by K. A. Youngstrom (1940. *J. comp. Neurol.*, **73**, 139–151). I am not sure exactly what they are supposed to do, but the secondary cells appear later in development. In the anurans, at least in some species, at metamorphosis the primary cells disappear entirely. In the salamander the primary and secondary cells are maintained throughout adult life. The axon of the secondary cell runs directly out from the spinal cord whereas the axon of the primary has a somewhat circuitous route and is usually a larger fibre. The secondary cell may make direct contact with the incoming afferent fibres, whereas there is an internuncial neuron interposed between the afferent endings and the primary cell.

Kollros: It was my impression that Youngstrom indicated that in the salamander the primaries are those which have both trunk innervation and a limb innervation, and the secondaries presumably just have limb innervation.

Piatt: The emphasis here must be on the word presumably. Youngstrom derived this idea largely from Coghill's earlier work.

Drachman: Dr. Székely, you have shown that limbs innervated by thoracic spinal cord will move only under two circumstances: (*a*) if more than one thoracic cord is made to innervate a single limb, or (*b*) if the grafted thoracic cord is in an early stage of development. It would seem that both of these procedures tend to increase the number of motor neurons available to innervate the limbs. In the first case the source of extra motor neurons is obviously the additional grafted cord. In the second case it is possible that a large number of presumptive motor neurons would be present in the spinal cord of a younger animal, as Dr. Hughes shows (this symposium). Would you accept this explanation of your findings?

Székely: Yes and no. The factor which determines limb movement is an increased number of nerve cells. But I doubt that in the case of very early cord transplantation this would be achieved by preserving all of the available neurons. If one makes the cord transplantation three to four stages later, just after a couple of hours delay, during which no cell loss can occur, limb movements do not develop. This suggests that the question belongs to the obscure problems of development: how can a "competent" grafted piece of thoracic cord be transformed into brachial cord by the influence of the surroundings? That I cannot answer.

GENERAL DISCUSSION

Eccles: We should try to clarify our ideas about the whole development of specific neuronal connexions. The nerve fibres grow out towards all kinds of possible targets and somehow or other they make connexions. We are thinking firstly in terms of Ramón y Cajal's neuronal theory, which everybody has assumed as the background. On the whole the targets are appropriate for the functional performance. Also I take it that no one is proposing action at a long distance—a millimetre or so. The clues the fibre has about where to go are then as follows:

(1) Free molecules of a specific kind diffusing from a surface so that the nerve fibre senses one kind of surface rather than another. That would work at most over a few microns.

(2) Specific surfaces—these of course would not be free-swimming molecules, but there can be quite complex steric patterns on surfaces which are fixed like receptor sites. Furthermore, it isn't just a matter of the surface being passive and the fibres active. The surface could also reject connexions. It could work both ways, and this mutual acceptance could be regarded as a purposive target of the outgrowing fibres, so that when the fibre makes some kind of contact the surface makes it permanent, or on the other hand it tries to unloose itself from this unwelcome embrace. We have to consider how highly specified the axons are as they grow out, how highly specified the targets are, and how the information gets across.

Szentágothai: Attractive forces acting from a distance have been discarded by P. Weiss (1943. *J. exp. Zool.*, **68**, 393–448); on the other hand ultrastructural organization of the intercellular matrix of the tissue spaces in which the fibres grow is obviously of crucial importance. There is, however, an additional important geometric factor that has to be taken into account. According to tissue culture experiments, nerve fibres tend to grow in straight lines, provided that the medium in which they grow has no macromolecular organization in a certain preferential direction. The direction of axon growth under these circumstances is determined by the initial direction that the axon has when starting to grow in this medium. Some ten years ago we investigated from this point of view (Szentágothai, J., and Székely, G. [1956]. *Acta biol. hung.*, **6**, 215–229) a number of axon developments, under both normal and experimental circumstances. It is well known from the histogenesis of the nervous system that axons start to grow not at random in any direction, but generally in a well-determined direction. This "primary axis orientation" of the neuroblast and the tendency of the axon to grow in straight lines would together secure some kind of "aiming" at a primary target. We have been able to show that in a number of cases (spinal ganglion, trigeminal ganglion, spinal motor neurons) the early orientation of the axon corresponds to the "simplest and straightest" way the axon takes during its further growth. The study of various degrees of experimentally induced cyclopean malformations of the eye has shown that the initial orientation of optic axons is decisive for the site of the chiasma formation. If there is no retinal coloboma, which guides the fibres away from the optic cup—as happens in complete Mangold type (Mangold, O. [1931]. *Ergebn. Biol.*, **7**, 193–403) cyclopean eyes—the chiasma is formed immediately behind the lens.

I do not want, of course, to introduce old-fashioned mechanical concepts, but in such a multi-factorial process as nerve fibre growth and target finding all possible aspects ought to be considered. Axonal growth—in certain aspects—might resemble a game of billiards, where the ball is aimed first in a certain direction and several further changes of direction have to be calculated. The axon, by the primary axis orientation of the neuroblast, might be aimed at a preliminary target, where this direction might receive a certain change guiding or re-directing it to a secondary target, etc.

Eccles: We could easily add the additional mechanical factors, correlating chemical information and mechanical guidance.

Székely: Paul Weiss used the term guiding surfaces (1950. *J. exp. Zool.*, **113**, 397–461) and he meant by this that the outgoing fibres can make their way along leading surfaces or along interfaces.

Eccles: Weiss called this contact guidance.

Sperry: Ever since Harrison (1914. *J. exp. Zool.*, **17**, 521) demonstrated it, it has been accepted that nerve growth is ubiquitously contact-guided. The only question is whether something additional like chemical selectivity is involved. Dr. Székely mentioned a general point in his review which involves two issues: one is whether the fibres grow out preferentially to get to their muscles, or whether the muscles specify the fibres, so that one ends up with nice connexions. Either way it is the same functionally, and both presume an equal order of biochemical specificity.

A much bigger issue which arose during the first 20 years of this work was Weiss's contention that *no* connexion system at all can explain the behavioural results. From the standpoint of neurophysiology the critical question is whether the nervous system works in terms of connexions. Weiss (1936. *Biol. Rev.*, **11**, 494) for many years had claimed that a single motor neuron growing out and connecting by branches to five different asynchronous muscles would fire each one properly and separately over the single axon such that no connexion system can explain this myotypic response. This is the old Resonance Principle of impulse specificity. Recently Dr. Székely seems to be reviving this old signal-specificity concept, denying connexions and talking in terms of impulse and pattern specificity. There is a curious history here.

Hughes: Time sequences in development play a very important part and I shall go into that tomorrow in relation to limbs.

Székely: I think, Professor Sperry, that a clear distinction is to be made between Weiss's concept of resonance theory and the specificity of impulse patterns (though I never used this term) I have discussed (1966. *Adv. Morphogen.*, **5**, 181–219). While in his early papers Weiss (1924. *Arch. EntwMech. Org.*, **102**, 635–672) maintained that a muscle could respond only to specific frequencies, I am trying to find a physiological background for some elective responses of the centre to different kinds of "compound" impulses without assuming a rigid specific connexion pattern among the nerve cells involved.

Sperry: He moved beyond that in the late 1930s to the idea of chemical sensitization, and I think you were suggesting frequency just two years ago.

Székely: No, I never suggested this in connexion with limb movements.

Hughes: I think there is a distinction between the ideas put forward by Dr. Weiss and those put forward by Dr. Székely. Weiss talked in terms of single connexions.

Hamburger: How do you envisage the role of specific cell surfaces in the formation of nerve pathways, Professor Eccles?

Eccles: The surface attracts the fibre, the fibre feels the attraction of the surface and therefore goes there until the surface reacts to the fibre and either accepts or rejects it. I take it that the cells have the propensity to reject an unwanted fibre even if the fibre wants to make contact with it. This again is something one can perhaps see in tissue culture.

Hamburger: An idea which also stems from Paul Weiss is that the matrix, i.e. the intercellular micellar arrangement of macromolecules, may play a role in the formation of nerve pathways. I would conceive of this as exudates of cells which may have the same kind of surface specificity which you attribute to the cell surfaces, Professor Eccles. It need not necessarily be cell surfaces which exercise this kind of selectivity—it could be extracellular exudates, which may be very highly specific.

Eccles: Of course many of these connexions are established over quite small distances and later the cells migrate and leave their synaptic connexions behind them, gradually stringing out their axons. This kind of thing is seen with the granule cells in the cerebellum.

THE BEGINNINGS OF CO-ORDINATED MOVEMENTS IN THE CHICK EMBRYO

V. Hamburger

Department of Biology, Washington University, St. Louis, Missouri

SPONTANEOUS ACTIVITY

When we started our investigation of motility in the chick embryo several years ago, it was our intention to correlate neurogenetic development with behavioural development, and furthermore to obtain information on the embryonic origins of the behaviour found after hatching. Right at the beginning of our study the confrontation with the actual performance of the chick embryo forced us to revise most of the previous interpretations of embryonic motility and to approach the problem from a new and different viewpoint. I shall not give a historical survey of observations and concepts here (see Hamburger, 1963), but shall focus immediately on what we consider the essential feature of embryonic motility in the chick, namely that from its beginning at three and a half days to at least 17 days, embryonic activity is the result of spontaneous, self-generated discharges of the nervous system that are independent of sensory input. We define "spontaneous" with Bullock and Horridge as "repetitive change of state of neurons without change of state of the effective environment; that is, activity without stimulation other than the standing conditions" (1965, p. 314).

There is no disagreement among observers about the origin and progressive elaboration of the pattern of motility. Beginning with a slight bending of the head by contraction of the neck muscles at three and a half days, motility spreads caudally to include the trunk and tail. Wing and leg movements begin at six and a half days, and movements of the beak, lower eyelid and eyeball are added to the repertoire at later stages (review in Corner and Bot, 1967). However, the initiation and nature of these movements has been a matter of controversy. Several investigators (e.g. Kuo, 1932) have considered them as reflexogenic. Against this claim stands the old observation of the eminent German physiological psychologist, Preyer

(1885), that the embryo is refractory to exteroceptive stimulation up to seven and a half days. This observation, which was confirmed by us and others, refutes the reflexogenic nature of motility up to that stage. Since the motility pattern does not undergo detectable changes after the structural basis for sensory input has been established, in all probability the greater part of motility after eight days and at least up to 17 days continues to be spontaneous and not guided by sensory information. Two characteristics of motility give strong support to this notion.

(1) Motility is periodic, phases of activity alternating with phases of inactivity. We have obtained quantitative data on this periodicity by mechanical recordings. The overall activity, which we define as the percentage of activity during a standard observation period of 15 minutes, increases gradually until a peak of 80 per cent is reached at 13 days. This peak is maintained until 17 days, whereupon activity declines. The duration of active phases increases and that of inactive phases decreases during the rise of activity (Hamburger et al., 1965). Rhythmic performance in itself suggests the spontaneous origin of motility.

(2) The movements are jerky random movements, including wriggles of the body, kicking of the legs, opening and closing of the lower eyelid, beak gaping, and in later stages, beak clapping. During an active phase, the different parts move independently of each other in constantly changing combinations which appear unpredictable. All parts that are capable of moving at a given stage participate during an active phase. However, the movements of the different parts are not coordinated with each other; for instance, the wings do not move together as in flight (though in later stages synchronous wing flapping was observed), nor do the legs move in an alternating pattern, as in walking. This type of disorganized movement is hardly compatible with the notion of control or guidance by sensory information.

A crucial test of the role of sensory input was provided by the following experiment. The lumbar region of the embryo was deafferented totally by a double operation performed on two-day-old embryos: extirpation of the dorsal half of the lumbar spinal cord including the precursor (neural crest) cells of spinal ganglia; and simultaneous extirpation of the entire spinal cord at the thoracic level, to exclude sensory input from rostral levels. In control embryos, only the thoracic gap was made. Percentage activity during our standard observation period of 15 minutes was the same in experimental and control embryos at least up to 15 days. A decline of motility between 15 and 17 days in experimental embryos can be attributed to progressive deterioration of the motor neurons. Spina bifida of the lumbar cord is the rule in these preparations. In careful histological checks, degeneration and loss of motor neurons was found to parallel closely the decline of motility (Hamburger, Wenger and Oppenheim, 1966). Incidentally, the control embryos as well as embryos with cervical gaps in the spinal cord show that isolated parts of the spinal cord are capable of periodic discharges. However, the percentage activity of such preparations is lower than that of intact embryos. We conclude that brain cells transmit discharges to the spinal cord, thus increasing the activity of the latter. Early embryonic extirpations of the medulla and cerebellum, respectively, have shown that we are dealing with specific influences from different brain regions which appear at different stages of development (Decker and Hamburger, 1967).

The chick embryo turns out to be a particularly favourable subject for this kind of investigation, since it demonstrates the potentials of its nervous system directly by its overt activity. The motility is probably the expression of the gradual maturation of the nervous system; neurons begin to generate discharges when they have attained a certain stage of differentiation. Incidental to this process is an effect that perhaps has biological significance. We know that temporary immobilization results in ankylosis (Drachman and Coulombre, 1962; Drachman and Sokoloff, 1966); and permanent deprivation of limb innervation results in the degeneration of all limb muscles (Eastlick, 1943; Drachman, 1964). It is possible that these pathological conditions are prevented by the spontaneous activities of the embryo.

HATCHING

The type of unorganized motility already described is hardly suitable for the escape of the embryo from the shell. In fact, it is known that the shell is cracked by a sequence of vigorous back thrusts of the beak, combined with a rotation of the whole embryo whereby the circumference of the blunt pole near the margin of the air-space is encircled. These movements have the appearance of a coordinated behaviour pattern. A detailed investigation has shown that all parts of the body are involved in this performance with the exception of the wings. Each thrust of the beak is accompanied by a bracing of the shoulder and of the intertarsal joints against the shell, and a wave-like writhing movement of the whole body. Some time after the inception of the beak thrusts, which gradually become more regular, a rotatory component is added in which the body and the tarsal joints participate. As a result, shell pieces are chipped off along an irregular circle near the blunt pole. When the embryo has rotated approximately two-thirds of the circumference, the cap is sufficiently loosened to be lifted off by vigorous thrusts of the shoulder and tarsal joints against the shell and powerful struggling of the legs and wriggling of the body (Hamburger and Oppenheim, 1967). We have referred to the hatching act as the climax.

COORDINATED PRE-HATCHING MOVEMENTS

Hatching cannot be accomplished successfully unless it starts from a characteristic position which the embryo attains usually 24 hours or more before the climax. We refer to it as the hatching position. The embryo is oriented lengthwise in the shell with the head near the blunt end. The neck is twisted in a tight coil. The right side of the head is tucked under the right wing. The beak has pierced the membrane which separates the embryo from the air-space and is poised obliquely against the shell. The trunk and tarso-metatarsal joints are closely applied to the shell. Significant deviations from the hatching position reduce the chance of successful hatching considerably. Again, it seems hardly conceivable that the undirected random movements characteristic of embryonic motility up to 17 days could lead to the stereotyped hatching position. We have made a careful study of movements between 17 and 19 days and have found a pattern of co-ordinated movements which bring the embryo into the hatching position through a sequence of intermediate stages. These movements are different from the climax pattern, but all pre-hatching movements have some features in common with the climax pattern.

At the beginning of day 17, the embryo is already oriented lengthwise in the shell; the neck is not yet twisted but is bent straight downward and the beak is buried in the yolk sac. The first directed movements, during day 17, lift the beak out of the yolk sac and twist the neck in such a way that the head is turned to the right and the tip of the beak approaches the right wing. Actually, the early rotatory head movements may be either to the left or to the right, but the movements to the right prevail eventually. Then follows a particularly interesting phase, the tucking of the beak and head under the right wing. This feat is accomplished by a sequence of thrusts of the head and beak towards the wing; these thrusts are more or less well synchronized with a rotatory lifting of both wings. During one of these simultaneous beak thrusts and wing-raising movements the beak slips under the wing and disappears from sight momentarily. Shortly thereafter, the tip of the beak appears behind the elbow and tucking is completed. However, one would get an erroneous impression if one visualized the process of twisting of the neck and the subsequent tucking manoeuvre as a well-executed, straightforward and direct action. Actually these procedures are drawn out over many hours and characterized by repeated reversals, false starts and deviations. All embryos seem to go through several tucking and untucking episodes before an irreversible position is maintained. However, when this is finally accomplished, the embryo is not yet in the hatching position. It takes another forward shift of the head towards the air-space to bring the beak near the membrane so that it can pierce it. This shift is again executed in many small steps. Each body wriggle and head movement has a clearly discernible rotatory component.

The pre-tucking, tucking and post-tucking procedures occupy the embryo during most of days 17 and 18. During day 19, the beak penetrates the membrane. This is accomplished by scraping of the closed beak, slow opening, whereby the membrane is drawn taut between the upper and lower beak, and by occasional sharp thrusts of the beak. These different movements go on for hours, until the membrane is worn thin and eventually ruptures. The first cracking of the shell ("pipping"), which may occur at any time between 24 hours and a few minutes before the onset of the climax, is achieved by the same type of movements that are observed during the climax; it is actually a brief and precocious climax episode. (For details of pre-hatching motility, see Hamburger and Oppenheim, 1967.)

CONCLUDING REMARKS

It is evident from our account that, beginning with day 17, complex coordinated movements are performed which are instrumental in the lifting of the beak out of the yolk sac, in the pre-tucking and tucking behaviour, in the attainment of the hatching position, in pipping and hatching. Although each phase has its own characteristics, all these activity patterns have several features in common; perhaps they can be considered as variations of a basic theme adapted to the specific goals which each of these activities has to attain. Among the common features are the involvement of the whole body and legs, and a rotatory component which is noticeable, distinctly and separately, in the head, trunk and wings.

It seems to us that these coordinated patterns of activity are fundamentally different from the random movements which are the only type of motility seen up to 16 or 17 days. The main difference between the two motility patterns is the directional character of the former and the non-directional character of the latter. Furthermore, the rotatory component was never observed in the random movement; nor is the bracing of shoulder and tarsal joints against the shell included in the repertoire of random movements.

The question arises whether the coordinated movements emerge from the random movements by organization and structuring of the latter. We believe that this is not the case; rather, the coordinated movements arise *de novo* at the appropriate stages of development. A strong argument in favour of this assumption is our observation that the random movements are merely suspended when the coordinated movements are performed, but resumed during the intervals. For instance, they are performed for hours between membrane penetration and pipping, and between pipping and the climax. It seems that periodic random movements continue to be performed after hatching (Corner and Bot, 1967).

Of course, the coordinated pattern of activity requires the establishment of centres of integration in the central nervous system. These centres probably differentiate and mature in earlier stages and are activated by endogenous or exogenous stimuli. At present, we are engaged in an experimental analysis of the role played by sensory information in the pre-hatching and hatching behaviour.

SUMMARY

Two different motility patterns were observed in the chick embryo. From the beginning of motility at three and half days to approximately 17 days, motility consists of unorganized random movements which are performed periodically, phases of activity alternating with phases of inactivity. During an activity phase, all parts of the body that are capable of moving at a given stage of development are involved; one observes jerky leg movements, trunk and tail wriggles and in later stages beak gaping and clapping and opening and closing of the eyelids. However, the movements of different parts, for instance those of the legs, are not coordinated with each other.

In contrast, the hatching process necessitates the coordination of all parts of the body. Oblique back thrusts of the beak against the shell are coordinated with vigorous writhing movements of the body and bracing of the shoulder and of the intertarsal joints against the shell. Body and leg movements include a rotatory component; as a result, the beak thrusts encircle the blunt pole near the margin of the air-space, chipping off pieces of shell along this line. Other types of coordinated movements are performed, beginning with day 17. They bring the embryo into the hatching position. Significant phases in the later process are: the twisting of the neck, whereby the beak is brought in front of the right wing; the tucking of the head under the right wing; the shift of head and body towards the airspace; and pipping of the shell.

It appears that the random movements and the coordinated movements are not related to each other. The former are suspended while the latter are performed, but resumed when the coordinated movements cease.

REFERENCES

BULLOCK, T. H., and HORRIDGE, G. A. (1965). *Structure and Function in the Nervous Systems of Invertebrates*. San Francisco and London: Freeman.
CORNER, M. A., and BOT, A. P. C. (1967). *Brain Res., Amst.*, **6**, in press.
DECKER, J. D., and HAMBURGER, V. (1967). *J. exp. Zool.*, **165**, 371-384
DRACHMAN, D. B. (1964). *Science*, **145**, 719-721.
DRACHMAN, D. B., and COULOMBRE, A. J. (1962). *Lancet*, **2**, 523-526.
DRACHMAN, D. B., and SOKOLOFF, L. (1966). *Devl Biol.*, **14**, 401-420.
EASTLICK, H. L. (1943). *J. exp. Zool.*, **93**, 27-49.
HAMBURGER, V. (1963). *Q. Rev. Biol.*, **38**, 352-365.
HAMBURGER, V., BALABAN, M., WENGER, E., and OPPENHEIM, R. (1965). *J. exp. Zool.*, **159**, 1-14.
HAMBURGER, V., and OPPENHEIM, R. (1967). *J. exp. Zool.*, **166**, 171-204
HAMBURGER, V., WENGER, E., and OPPENHEIM, R. (1966). *J. exp. Zool.*, **162**, 133-160.
KUO, Z.-Y. (1932). *J. comp. Psychol.*, **13**, 245-271.
PREYER, W. (1885). *Specielle Physiologie des Embryo*. Leipzig: Grieben.

DISCUSSION

Hughes: Dr. A. Bellairs, Dr. S. Bryant and I have followed movement in the embryo of the European viviparous lizard. Exactly the same build-up of spontaneous activity rising to a plateau occurs as in the chick. What surprised us was that the sinuous axial motion, the most primitive movement of vertebrates, only appears right at the end of embryonic development, which is exactly what you find for purposive behaviour in the chick, Professor Hamburger.

Hamburger: In the chick one could speak of two types of sinuous movements. One type occurs at the beginning of motility, at four to five days. Wriggling of the head and body, with a rotatory component, occurs from 17 days on, preparatory to hatching.

Dr. J. Decker in our laboratory has found the same build-up of periodic spontaneous motility in turtle embryos. Motility declines after a peak is reached (1967. *Science*, **157**, 952-954).

Muntz: In the anuran (*Xenopus laevis*) I have seen similar spontaneous movement which increases from the beginning of the reflexogenic stage, that is when there are connexions between the sensory and motor nerves. This is the first time we seem to get any spontaneous movement. Earlier movements are in response to mechanical stimulation.

Hamburger: But in amphibians there is no period during which there is only motor innervation and no sensory innervation. According to G. E. Coghill (1929. *Anatomy and the Problem of Behaviour*. Cambridge University Press), both sensory and motor pathways are present when motility begins in salamander embryos. The very first movements can be elicited by stimulation of the skin.

Muntz: Xenopus laevis has sensory nerves in the form of Rohon-Beard cells which are not connected either centrally or peripherally at a stage when the primary motor nerves are only just beginning to differentiate. At the stage when the sensory and motor nerves are connected centrally and peripherally in *Xenopus*, spontaneous movements begin which increase until what would be the hatching stage, when it is capable of coordinated swimming. After hatching there appear to be fewer spontaneous contractions.

Szentágothai: Would you regard the movements leading to hatching in the chick embryo as some kind of goal-directed behaviour or as a very elementary stereotyped pattern of movements? Because of the way the embryo is accommodated in the egg space, such stereotyped movements may with some statistical probability lead to the head being lifted up and then put behind the wing. If this movement were coordinated with the movement of the right wing, one would not need to suppose any goal-directed behaviour, but it would be just a mechanical sequence of events that with a large statistical probability would lead to the final hatching position and to the piercing first of the membrane and then of the shell.

Hamburger: The term "goal-directed" is loaded and I would use it only with reservations. The movements described as coordinated are directed: they lead to an irreversible position, whereas the random movements have no preferential direction. Furthermore, in the coordinated movements the whole body is usually involved, there is a rotatory component, and there is bracing of the shoulder and of the tarsal joints against the shell. These three features have never appeared as part of the repertoire of undirected random movements. I use the term "coordinated" for movements in which there is coordination of head, body, leg, wing and feet movements in the performance of a complex action pattern, such as tucking or hatching. It is a very primitive type of coordinated movement which perhaps doesn't exist in this form in later post-hatching behaviour.

Szentágothai: I still would differentiate this primitive kind of coordinated movement from the kind seen when the chick is hatched. Immediately after hatching the chick shows what I call, although it may not be a good expression, "goal-directed behaviour". I just want to ascertain whether you think that these three levels can be distinguished.

Hamburger: I would be inclined to agree. For instance, it is conceivable that sensory input, which is obviously very important in "goal-directed" post-hatching behaviour, plays a less important, or a different, role in pre-hatching and hatching behaviour. Sensory input is not involved in the uncoordinated random movements.

Walton: Is the pressure of the shell a necessary stimulus? What happens if you remove the shell and leave the embryo in its membrane?

Hamburger: That is very difficult. I have not accomplished it yet.

Kerkut: There should be some sensory phenomena coming in, because M. A. Vince (1966. *Anim. Behav.*, **14**, 34–40) has shown that embryos within the egg are making little noises, and that an isolated egg does not hatch out as quickly as one where other eggs are making noises.

Hamburger: There is no doubt that an elaborate sensory machinery is present during the pre-hatching and hatching period. The role it plays remains to be analysed. Experimental tests of the role of the vestibular system and of the trigeminal sensory input are being made in our laboratory.

Kerkut: It speeds up the hatching process.

Hamburger: Perhaps. We don't know anything about the triggering of these pre-hatching and hatching action patterns. All I know is that they are poorly programmed but coordinated.

Eccles: If you admit that there are sensory inputs going into the organism, and you are still dubious about whether it alters the motor output, then you would disagree with neurophysiologists on the whole. We do believe there is a hook-up between sensory inputs and motor outputs. However, it may not be the sole determining factor in the motor output.

Kollros: You indicated that the chick embryo was fairly quiescent at 17 to 19 days, and that there were many degenerating motor cells. Are the degenerating cells originally connected with the periphery?

Hamburger: Yes. Up to 15 days all the experimental embryos show normal motility and a normal motor innervation.

Eccles: That does not quite answer the question because only some would need to be connected, and that we don't know.

Kollros: Does a new generation of cells develop?

Hamburger: No, because the mitotic activity could not be revived.

Buller: Could little chickens come out of big shells? This might answer Professor Szentágothai, who is saying that it is space that matters and it is therefore a packing problem.

Hamburger: I don't know how to get smaller embryos into the shell. One could perhaps remove the yolk.

Buller: Is there a fixed relation between the volume of the egg and the weight of the chick that comes out?

Hamburger: I always find that the chick fills its space completely, apart from the air chamber.

Drachman: I have observed tucking of the head under the right wing in some of the chick embryos which have been subjected to prolonged total paralysis in our laboratory. Do you think that under these circumstances the growth of the embryo within the confines of the shell may lead to this posture?

As you know, we have studied the nature of embryonic movements from a pharmacological point of view. We found (Drachman, D. B. [1965]. *J. Physiol., Lond.*, **180**, 735–740) that chick embryos which are continuously paralysed with curare for as long as 24 hours resume some muscular contractions which are not abolished by even very large additional doses of curare. One possible explanation of this phenomenon is that the resumed movements may be primarily myogenic. That is, the impulses may originate in muscle rather than being transmitted from motor nerves. Do you think that at the earlier ages some of the spontaneous embryonic movements may normally be myogenic in origin?

Hamburger: We have data only for $3\frac{1}{2}$- and 4-day embryos, that is, the very beginning of neurogenic motility. There is no myogenic phase in the chick embryo. Mrs. B. B. Alconero (1965. *J. Embryol. exp. Morph.*, **13**, 255–266) has shown this by isolating somites with spinal cord and somites without spinal cord and then transplanting these structures onto the chorioallantoic membrane. The preparations which include the spinal cord show spontaneous periodic motility, whereas somites without spinal cord never show any sign of movement, although the muscles are perfectly normally differentiated.

Székely: Did you stimulate the group of muscles without nerves?

Hamburger: If these somite muscles are stimulated very strongly contraction is seen.

Székely: That is myogenic.

Hamburger: I agree.

Eccles: Muscles do exhibit spontaneous mechanical activity, probably even before they are innervated (Diamond, J., and Miledi, R. [1962]. *J. Physiol., Lond.*, **162**, 393–408).

Hník: What is known about the sensory innervation of muscles and skin in these embryos? At what time do sensory nerve fibres grow into the tissues?

Hamburger: The skin is innervated at seven to eight days; proprioceptive muscle sensitivity begins slightly later.

Levi-Montalcini: We do not know when a first provisional contact is established between growing motor nerve fibres and muscles, nor do we know when such temporary contact is replaced by a permanent one.

Hamburger: Maybe at 12 to 13 days muscle spindles are formed.

Hník: You showed that uncoordinated activity suddenly increased at about the 14th day. Could this be due to the accessory influence of sensory innervation?

Hamburger: I think the sudden rise is due to an increased number of neurons that begin to fire and activate the motor neurons.

Eccles: So you are proposing that this spontaneous activity is not just motor neurons firing impulses but that it could be interneurons firing motor neurons which then fire the muscles. But I am not at all averse to the idea that motor neurons also may be operating spontaneously. This problem is quite open, I think, as to how the neuronal machinery of the spinal cord is generating spontaneous movements.

Hamburger: We are trying to record from different parts of the spinal cord.

Mugnaini: Coordinated motility requires maturity of various centres in the central nervous system. Nidifugous birds at hatching have a quite mature CNS, cerebellum included, while nidicolous species are in general neurologically more immature. It could therefore be interesting to analyse the CNS of nidicolous birds at hatching, to see whether there are centres presenting a relatively more advanced degree of differentiation, which could be related to the hatching movements.

Hamburger: I would predict that those centres which are necessary have matured in advance of the event. It would be extremely interesting to make comparative studies.

[Professor Hamburger then showed a film, made by Dr. J. Decker, of the hatching process of a chick embryo.]

DEVELOPMENT OF LIMB INNERVATION

A. Hughes

Department of Zoology, University of Bristol

For 60 years it has been known that the nerve fibre is formed as an outgrowth from a neuroblastic cell. In his pioneer experiments on the cultivation of tissues *in vitro*, Harrison (1910) showed that an embryonic nerve fibre can extend through a medium of clotted fibrin, and so does not necessarily depend on any local protoplasmic accretions from its surroundings. By inference, the same is true for the outgrowing fibre within the intact organism.

This demonstration left on one side the problem of how the nerve axon finds its way through the tissues of the embryo. Indeed, the question of the determination of nervous pathways thereby became an even deeper mystery. It was shown that nerve fibres tend to grow along surfaces (Harrison, 1914), yet to postulate that the whole pattern of the nervous system is preceded by a complete system of pre-formed guiding tracts (Weiss, 1941) is to re-state the problem in terms of entities for which no direct evidence has yet been produced. Moreover, whatever the guiding principles are, the growing nerve fibre often fails to conform (Ramón y Cajal, 1908).

Nearly all the experimental evidence concerning the directional growth of nerve fibres within the organism is concerned with nerve regeneration. Here two apparently separate concepts have been put forward. Researches in lower vertebrates on the regeneration of the optic nerve after section have clearly shown that fibres growing once more from the ganglion layer of the retina towards the brain finally reach the optic tectum in the ordered arrangement of terminals necessary to re-establish vision (Attardi and Sperry, 1963), even though at the site of the division of the optic nerve they traverse scar tissue in a wholly random fashion (Sperry, 1951). It thus seems that the regenerating fibre is from the first endowed with some property by which it can "find" its appropriate end-station within the brain.

This highly determinate behaviour is in apparent contrast to the results obtained by grafting supernumerary limbs in the tailed Amphibia. If a grafted limb is made to share the nerve supply of a nearby member through a common limb plexus, it is later found that the two limbs move in harmony, even though the anatomical pattern of the nerves regenerated within the graft differs from the normal arrangement (Detwiler, 1925; Weiss, 1922, 1937). This "homologous response" extends to the individual muscles within each limb and is seen where the supernumerary member is transplanted in a reversed orientation, or even to instances where a single muscle is grafted and becomes innervated at random from a local supply (Weiss, 1931). Weiss's explanation of these remarkable results was that the regenerating fibres enter the graft and establish contact with muscles wholly at random, and that furthermore the muscle itself exercises a controlling influence upon the development of functional relationships on the neurons with which it comes into contact. This extends even to a selective acquisition of synapses round each motor cell body within the cord. Thus the central pattern of coordinating relationships is derived "myotypically" (Weiss, 1941) in response to directive influences emanating from the periphery. The outgrowing fibres which approach the grafted limb are in the first place wholly undetermined, with respect both to destination and to their ultimate function.

Within recent years however, experiments at early stages of development on the interchange of lengths of spinal cord between brachial and lumbar regions have shown that the behaviour pattern of a limb is determined from the centre. In a post-embryonic chick the movement of a wing innervated by lumbar cord resembles that of a leg (Székely, 1966). This demonstration of local factors within the cord which shape central patterns for the coordination of limb movement demands some answer to the problem of how the nervous circuits linking cord and limb are achieved in the first place. Do the first fibres of the embryonic spinal roots find their way directly to their final end-stations, or is there a random element in their early outgrowth?

One principle which has been shown to operate in the progress of innervation of the limb is a proximo-distal order of growth which is seen not only in the early phases of limb formation (Saunders, 1948; Tschumi, 1957), but also at later stages of development (Hughes and Prestige, 1967). Thus the first movements of the limb are at the hip, while motility at other joints follows later (Hooker, 1952). Furthermore, the proximo-distal development of limb innervation is linked with a cranio-caudal differentiation of the ventral horn within the cord.

In the mammal, the texture of the ventral horn changes during foetal life from a compact mass of differentiating neurons to one composed of separate columns of motor cells, each of which innervates a group of muscles of related function. Romanes (1941) has shown that in the rabbit foetus the separation of the cell columns occurs at the same time as a growth of axons toward limb muscles. The first columns of cells to become distinct are at the cranial pole of the ventral horn, and from them axons recognizable in silver-stained preparations grow towards their related groups of myoblasts within the thigh.

These developments continue until the more caudally situated motor neurons have dispatched bundles of axons to sites in distal segments of the limb. Thus in the developing mammal there is evidence of a directed outgrowth of nerve fibres into the limb. Furthermore, it is possible that this may be controlled by time sequences of histogenetic processes within ventral horn and limb. One may argue that while factors within the ventral horn cause axons to grow at the same time as their cell bodies cluster together in columns, other factors simultaneously ensure their reception by groups of myoblasts within the limb. It seems unlikely, however, that any similar sequential relationships operate in the regeneration of fibres of the optic nerve (Gaze, 1960).

In the developing anuran, the anatomical pattern of the future limb nerves is sketched out by the first fibres to enter the limb bud (Taylor, 1943). These fibres are relatively few and are greatly outnumbered by later arrivals. There is evidence from the reinnervation of grafted limbs in the embryonic anuran *Eleutherodactylus* that these early limb fibres have specially invasive properties (Hughes, 1962, 1967).

Although we lack any direct demonstration of serial innervation of the amphibian limb such as Romanes' observations have provided for the mammal, there is evidence that at early stages of limb motility, thigh muscles are innervated by more mature neurons than are those of more distal segments. This evidence comes from study of the changes during development in the immediate reactions of ventral horn cells to axotomy consequent on the amputation of a limb. In *Eleutherodactylus* there is a sharp change at eight days of development from a phase in which amputation results in widespread degeneration of ventral horn cells within 24 hours of operation to one in which this loss of cells is postponed for several days (Hughes, 1967). During this second phase of reaction, the normal turnover of cells within the ventral horn accompanied by a decline in total numbers is suspended as a result of axotomy, with the result that on the amputated side, more cells are counted temporarily than in the contralateral horn.

If, at this transition point, one compares the effect of amputation at the hip and at the knee, it is seen that the immediate decrease in total number of cells after the latter operation is much the greater, for when the thigh is ablated along with the other limb segments, the cessation of cell turnover among the more mature motor cells which innervated thigh muscles temporarily compensates in numbers for the loss of less differentiated cells which supplied more distal segments. When similar operations are performed two days later at a time when ventral horn cells in all regions of the limb respond to axotomy in this second phase of reaction, little difference is then seen between embryos amputated at either level.

A rough regional distribution of motor axons within the developing anuran limb is thus established by the time that its first movements are evident. This pattern undergoes further elaboration and refinement, as is evident from the results of stimulating the individual lumbar nerves at each subsequent stage of development.

In the mature anuran, part of the response to stimulation of the most anterior lumbar nerve (S8) is flexion, largely at the hip and knee, while with the caudal members of the series (mainly S10), limb extension results, chiefly of the distal limb segments (Sherrington, 1892; Hughes and Prestige, 1967). In juvenile and adult individuals of *Xenopus laevis*, where movements of the hindlimb are largely confined to the horizontal plane, this general pattern is seen at its simplest. This pattern of response to the stimulation of limb nerves emerges only gradually during development, and at first there results a wide range of movements among which the definitive reactions are at first absent or inconspicuous. Thus at a stage when the normal movement of the limb is confined to flexion at the hip joint, stimuli applied to S8 may result in random and uncoordinated trembling in all segments, or in a general tetanic contraction throughout the limb. With S10 the final pattern of response appears at a later stage than with S8. We thus have evidence of selection of final nervous pathways within an antecedent pool of random connexions.

Is there any histological evidence for the operation of these postulated selective mechanisms? It has already been mentioned that within the differentiating anuran ventral horn there is a process of cell turnover. Fresh cells are continuously recruited from the adjacent mantle layer of the cord, while many within the ventral horn undergo degeneration. The net result of these changes is that the total number of ventral horn cells falls, in *Xenopus* from an initial total of about 5,000 to a final figure of 1,200 to 1,800 (Hughes, 1961). This decline is steepest at the time when degeneration is most conspicuous. Yet estimates of the total number of cells which degenerate within the ventral horn during larval life suggest that far more cells are lost than finally mature into motor neurons. Some process of selection may thus be postulated, though the mechanisms of control are still obscure. It is of possible significance that the relatively short period of the peak in cell turnover precisely coincides with the time when the final pattern of response to stimulation of the limb nerves is emerging. If these contemporaneous events are causally related, this would imply that among the cells of the ventral horn which degenerate would be those whose axons had formed connexions with muscle fibres incompatible with the final and definitive pattern. There is evidence that some of the ventral horn cells which degenerate had already sent axons into the lumbar ventral roots.

The thesis that the final pattern of limb innervation in the Anura is derived by a gradual series of developmental changes receives support from pharmacological evidence (Hughes, 1965). In a fully functioning neuromuscular system, not only are opposed groups of muscles separately innervated, but the action of each is accompanied by the inhibition of the motor cells which innervate its antagonist through, as is generally thought, special inhibitory synapses with their own transmitter substance (Eccles, 1957, 1964). This mechanism is itself inhibited by treatment with strychnine, and although some aspects of the pharmacology of nervous inhibition are still being debated (Eccles, Schmidt and Willis, 1963; Kellerth, 1965), the effects of strychnine on the developing neuromuscular system give some indication of the stages when inhibitory circuits are present. Thus axial swimming movement in larvae of *Xenopus* is sensitive from early stages, while the drug has no effect on the first movements of the limbs. Not until the appearance of the extensor thrust are they thrown into tetanus by strychnine; this is true both in the larvae of *Xenopus* and in *Eleutherodactylus* embryos.

Further information concerning changes within developing peripheral nerves can be obtained by counting their constituent fibres (Hughes, 1965; Prestige, 1967). The errors involved in quantitative histological work of this kind are considerable, but are much smaller than changes during development which occur in the numbers of fibres which can be counted in well-silvered sections of the spinal roots, though the light microscope gives no information about any axons less than half a micron in diameter which may be present. Within these limitations, study of the spinal roots and peripheral nerves during development in both *Eleutherodactylus* and *Xenopus* shows a general decrease in the number of constituent fibres in the late embryo.

In *Eleutherodactylus* there is a steep decline during the last two days before the hatching period; at the periphery the losses within the larger nerves of supply halve their numbers of fibres. Comparison of the changes which occur at various levels within the limb nerves suggests a decrease in the extent of branching of nerve fibres.

Further evidence in *Eleutherodactylus* of the withdrawal of axonal branches during later stages of development comes from studies of the innervation of grafted forelimbs which, when innervated after transplantation, become motile (Hughes, 1964). Movement in such grafts can thus be used as an indication during life of the presence of a motor nerve supply. When such limbs are transplanted at five or six days of development to a site near an undisturbed hindlimb, some receive a small number of collateral branches from the already established supply to the nearby member, into which axons from dorsal and ventral roots have already penetrated.

Towards the end of embryonic life it has been found that such grafts lose their motility, and that this loss is correlated with a retreat of their motor innervation, as judged by the comparison of batches of embryos fixed at different times after grafting at the same stage of development. In contrast to these results, a forelimb grafted in place of a hindlimb is innervated directly by regenerated axons, which persist into juvenile life with the retention of mobility in the graft.

It may be that a decrease at late stages of development in the incidence of branching normally plays a part in defining and giving further precision to the pattern of distribution of fibres within peripheral nerves. A hint that in the late human foetus fibres are lost within peripheral nerves is given by the observation of Gamble (1966) of degenerating axons within the ulnar nerve. In early post-foetal life, counts of fibres in peripheral nerves again increase in number (Agduhr, 1920), though it is uncertain whether this is due to the differentiation of fresh neurons in cord and ganglia, or to growth in diameter of fine fibres hitherto invisible in silvered preparations.

Another change related to functional development among the axons of peripheral nerves is their differentiation into several categories by size. This may precede any loss in total numbers, as in the nerve to the triceps femoris muscle in *Eleutherodactylus*, in which a few relatively large fibres appear as soon as the muscle takes part in the extensor reflex of the leg (Hughes, 1966). In embryos deprived of thyroid hormones, but in which normal limb motility develops, these larger fibres are evident at the same time as the extensor reflex of the leg, although the total number of fibres present is well below normal.

In the mammal, differentiation of fibre size in muscular nerves does not occur until post-foetal life, to judge from the results of Evans and Vizoso

(1951) concerning the nerve to the medial head of the gastrocnemius in the young rabbit, where the adult bimodal distribution of fibre sizes is not fully attained until the fourth month, at which time the post-foetal increase in the total number of fibres within the nerve is still in progress.

In the adult rabbit regeneration of this nerve after division is followed both by loss of fibres within the peripheral stump and by the regaining of separate size categories. Aitkin, Sharman and Young (1947) showed that the number of fibres on the proximal side of the injury at first increases many times above the normal level, presumably by branching in response to axotomy. These fibres all remain small in calibre unless terminal contact with a muscle is achieved, when some gradually increase in diameter, while many others are withdrawn.

SUMMARY

The picture of the development of the innervation of the tetrapod limb which is here outlined is that the fibres which grow into the embryonic limb are regionally distributed from the beginning, in such a way that the proximal segment receives fibres mainly from the cranial region of the ventral horn, while the later-formed distal segments of the limb are correspondingly innervated from mainly younger and more caudally placed motor cells.

There is evidence that among the Anura this arrangement is subsequently made increasingly precise by the elimination of irrelevant nervous pathways through mechanisms of cell and fibre selection.

REFERENCES

AGDUHR, E. (1920). *J. Psychol. Neurol., Lpz.*, **25**, 463–626.
AITKIN, J. T., SHARMAN, M., and YOUNG, J. Z. (1947). *J. Anat.*, **81**, 1–22.
ATTARDI, D. G., and SPERRY, R. W. (1963). *Expl Neurol.*, **7**, 46–64.
DETWILER, S. R. (1925). *J. comp. Neurol.*, **38**, 461–490.
ECCLES, J. C. (1957). *The Physiology of Nerve Cells.* Baltimore: Johns Hopkins.
ECCLES, J. C. (1964). *The Physiology of Synapses.* Berlin: Springer.
ECCLES, J. C., SCHMIDT, R. F., and WILLIS, W. D. (1963). *J. Physiol., Lond.*, **161**, 283–297.
EVANS, D. H. L., and VIZOSO, A. D. (1951). *J. comp. Neurol.*, **95**, 429–461.
GAMBLE, H. J. (1966). *J. Anat.*, **100**, 487–501.
GAZE, R. M. (1960). *Int. Rev. Neurobiol.*, **2**, 1–40.
HARRISON, R. G. (1910). *J. exp. Zool.*, **9**, 787–848.
HARRISON, R. G. (1914). *J. exp. Zool.*, **17**, 521–544.
HOOKER, D. (1952). *The Prenatal Origin of Behaviour.* Lawrence, Kan.: University of Kansas Press.
HUGHES, A. (1961). *J. Embryol. exp. Morph.*, **9**, 269–284.
HUGHES, A. (1962). *J. Embryol. exp. Morph.*, **10**, 575–601.
HUGHES, A. (1964). *J. Embryol. exp. Morph.*, **12**, 27–41.
HUGHES, A. (1965). *J. Embryol. exp. Morph.*, **13**, 9–34.

Hughes, A. (1966). *J. Embryol. exp. Morph.*, **16**, 401–430.
Hughes, A. (1967). *Aspects of Neural Ontogeny.* London: Logos Press.
Hughes, A., and Prestige, M. C. (1967). *J. Zool. Lond.*, **152**, 347–359.
Kellerth, J.-O. (1965). *Acta physiol. scand.*, **63**, 469–471.
Prestige, M. C. (1967). *J. Embryol. exp. Morph.*, **17**, 453–471.
Ramón y Cajal, S. (1908). *Anat. Anz.*, **32**, 1–25, 65–87.
Romanes, G. J. (1941). *J. Anat.*, **76**, 112–130.
Saunders, J. W. (1948). *J. exp. Zool.*, **108**, 363–404.
Sherrington, C. S. (1892). *J. Physiol., Lond.*, **13**, 621–772.
Sperry, R. W. (1951). In *Mechanisms of Neural Maturation*, pp. 236–280, ed. Stevens, S. S. New York: Wiley.
Székely, G. (1966). *Adv. Morphogen.*, **5**, 181–219.
Taylor, A. C. (1943). *Anat. Rec.*, **87**, 379–413.
Tschumi, P. A. (1957). *J. Anat.*, **91**, 149–173.
Weiss, P. (1922). *Anz. Akad. Wiss. Wien*, **59**, 198–199.
Weiss, P. (1931). *Pflügers Arch. ges. Physiol.*, **226**, 600–658.
Weiss, P. (1937). *J. comp. Neurol.*, **66**, 481–535, 537–548.
Weiss, P. (1941). *Comp. Psychol. Monogr.*, **17**, 1–96.

DISCUSSION

Piatt: Why is the ectopic hindlimb not motile although the transplanted forelimb is motile?

Hughes: It is peculiar. I think Paul Weiss found the same in *Rana*. The hindlimb in the amphibian is innervated from several nerves, whereas the forelimb is innervated by only $1\frac{1}{2}$ nerves. In the course of a very long series of experiments, I have never seen a hindlimb in forelimb position become motile. If the hindlimb bud is cut off and immediately put back in its original place, it will usually become motile.

Piatt: But these forelimbs were more in the region of the hindlimb.

Hughes: Yes. A few fibres coming from one segment are enough to make a forelimb motile in *Eleutherodactylus*.

Piatt: Does segment 8 innervate the more proximal part of the limb, 9 a little further down, and 10 the more distal part?

Hughes: In broad outline that is true.

Piatt: Is this contrary to J. S. Nicholas and D. H. Barron's findings (1935. *J. comp. Neurol.*, **61**, 413–431)?

Hughes: This is in an anuran and it might be different in the urodele.

Kollros: We have worked with *Rana pipiens* so our findings differ a little from yours. In *Rana pipiens* larvae (with forelimb emergence at stage XX and the end of tail loss at stage XXV), we find likewise that the first appearance of the lateral motor column, as distinct from the adjacent grey mantle, is at stage V (Reynolds, W. A. [1963]. *J. exp. Zool.*, **153**, 237–250; Beaudoin, A. R. [1955]. *Anat. Rec.*, **121**, 81–95). We have the same kind of decline of motor nerve cell number, although perhaps somewhat more gradual, as you indicated, Dr. Hughes. It starts by stage VII and the major decline is at about stages XIII and XIV, which is very close in terms of the actual figures to the ones you showed there at the same time. We find about 5,000 cells at stage V and about 1,800 at stage XXV. In regeneration studies we never find more cells than this at one time, even at warm temperatures. If we keep hypophysectomized animals 6, 9 or 18 months the cell number remains at its initial high value, about 5,000. If we give these animals hormones, the limbs grow and the cell numbers decline (Kollros, J. J., and Race, J., Jr. [1960]. *Anat. Rec.*, **136**, 224; Race, J., Jr. [1961]. *Gen. comp. Endocr.*, **1**, 322–331). If we amputate the limb bud during the larval period at stages V or VI, we get a rapid cell loss on the operated side; later, cell loss on the normal side begins, accelerates, and by stages XI–XIII we have more lateral motor column cells (though smaller ones) on the side of amputation than on the control side (Pearson, C., and Kollros, J. J. [1961]. *Am. Zoologist*, **1**, 466). If, instead, we amputate one limb in the hypophysectomized animal, we retain the original cell numbers, so there is an anomalous retention of cell numbers.

We have been concerned with your statement that there is recruitment of cells during the period of regeneration, Dr. Hughes. One way of getting information on this point is to attempt to label cells with [^{3}H]thymidine during the period of their production and then follow them later. If this kind of recruitment occurs labelled cells should move into the column from the grey mantle. There is a problem as to when to label and so far we have labelled only at stages IV and V (unpublished work) and we find a maximum of 20 cells labelled out of about 1,800 at the end of metamorphosis. This suggests that if there is recruitment it is very small, or alternatively, that the production of cells for the recruitment occurs still earlier and they are in the process of migrating towards the horn when we attempt to label. We cannot yet distinguish between these two possibilities.

As regards the sequence of behavioural development, A. C. Taylor and I in 1946 (*Anat. Rec.*, **94**, 7–23) indicated the first movement of the limb at stage IX in *Rana pipiens*, which is exactly identical to stage 54 in *Xenopus* in appearance. D. G. Dunlap (1966. *J. Morph.*, **119**, 241–258) described the proximodistal sequence of muscle development in the hindlimb of the frog from the stage of the pre-muscle mesenchyme condensation onward. I believe cross-striations are first visible at the thigh at about stage IX, which would be about stage 54 in *Xenopus*, and then they progress on down the limb. So to what extent is motility of distal limb segments correlated with the differentiation of the muscles in the proximal to distal sequence in *Xenopus*?

Hughes: Dr. Muntz has found that in *Eleutherodactylus* the array of myosin and actin filaments is present from the earliest stages of limb motility, which is correlated with a surprising degree of maturity of muscle histogenesis.

Prestige: In *Xenopus* we haven't actually checked on the degree of striation in the muscles, but the order in which the muscles differentiate would fit very nicely with the order in which the limb members become motile, Dr. Kollros.

Gutmann: Do the neurons which are destined to degenerate form temporary neuromuscular contacts? The levator ani muscle of the rat, for instance, degenerates before birth in the female rat, but persists if testosterone is applied (Gutmann, E., Hanzlíková, V., and Čihak, R. [1967]. *Experientia*, **23**, 852). In this case also the neuron is retained. If no testosterone is applied to the female rat a loss of the nerve innervating the muscle is eventually observed. This finding is related to your information on the effect of hormones on loss and retention of motor neurons during development.

Hughes: I think there is turnover at every level—in the cell bodies, in peripheral nerves and perhaps in the muscles themselves.

Drachman: What is the precise state of the peripheral connexions of the cells which are degenerating? Do they contact muscle functionally or anatomically? Are there mature end-plates?

Hughes: At these stages the nerve-muscle contacts are of the simplest kind. There are just axons running alongside muscle fibres. Histochemical tests for cholinesterase are positive, but differentiated end-plates are still a long way away. As to the distinction between axons and neurons destined to degenerate, at present we know nothing.

Kerkut: Dr. Kollros, in your hypophysectomized animals when you gave hormones, did the neurons then disappear?

Kollros: Thyroid hormone makes them disappear, in the same way as they would normally. It changes the cell number in essentially the normal pattern, so that if we have a different stage of limb appearance we have the appropriate cell number for that stage. But we do not necessarily have the appropriate cell size.

Kerkut: Do you think the thyroid hormone is acting on all neurons as much as it does on the Mauthner cell, which shrinks?

Kollros: I shall talk about that in my paper.

Székely: Dr. Hughes, you said that hindlimbs put in the place of forelimbs do not move, but that grafted forelimbs in the place of hindlimbs do move. Which pattern of movement does the grafted forelimb show in your interesting animals? In the frog the forelimbs and the hindlimbs move differently.

Hughes: There are two general patterns of movement: first of all the bilateral—the jump to get out of the way—and secondly alternate movements. After strong stimuli the animal jumps bilaterally. Innervated forelimbs in place of hindlimbs always move as hindlimbs.

Székely: Does the musculature of the grafted hindlimb develop or degenerate?

Hughes: In *Eleutherodactylus* no grafting can be done before about four days. At that time it is too late to get enough nerve fibres into a hindlimb graft in the place of forelimb, so only a few muscles get innervated.

Székely: And how do the muscles of a grafted hindlimb differentiate?

Hughes: Generally speaking the hindlimb in place of the forelimb doesn't develop at the normal pace, and there is a good deal of muscle wastage.

Eccles: This loss of motor neurons that have already had axonal outgrowth, as Dr. Kollros has found also, is certainly an extraordinary finding.

Kollros: The numbers of cells in the lateral motor column do decrease. I do not know of any evidence connecting a particular developed neuron with a particular nerve fibre in the periphery. In relation to Dr. Drachman's question, I think the best findings we have on the correlation between cell number and fibre number are given both by Dr. Hughes in *Xenopus* or *Eleutherodactylus* and by J. M. Van Stone (1964. *J. exp. Zool.*, **155**, 293–302) in *Rana sylvatica* or other frogs, where there are, let's say, 5,000 motor nerve cells to begin with, plus a substantial number of sensory cells, and no more than 1,600 or 1,800 fibres are found in the sciatic nerve at any time. In other words not all of those cells can possibly have sent their fibres to the periphery.

Eccles: And the inference is that the degenerating ones are those which failed to get their fibres to the periphery. Dr. Hughes stated that even when fibres grew to the peripheral nerves their parent cells were eventually destined to die.

Hughes: The only evidence for that is from adding up numbers of cells and fibres.

Eccles: But on a number count it looks as if some at least already had a peripheral action. There are all kinds of ways in which the nervous system gets put together. This is a quite basic idea in our whole conference. One is apt to think always that this happens only in a positive way—more and more cells being connected to the right places. But there can quite well be a lot of cutting away of neurons that have made unfavourable and inappropriate connexions. Is there a signal back of an appropriate kind? How do the successful neurons know whether the other ones survive?

Prestige: In *Xenopus* the total number of ventral horn cells starts at around 5,000 and goes down to under 2,000. Amputation of the leg in a juvenile has no effect on the number of ventral horn cells for four or five months and some cells don't finally degenerate until about eight months later. All the cells in the ventral horn of the juvenile are therefore very resistant to amputation, though not totally so. If you amputate at earlier stages of development, the cells are less resistant to amputation, and at the digit stage nearly all cells will degenerate immediately. Before this stage though, motor neurons are independent of the periphery and unaffected by amputation. Thus motor neurons develop from a phase in which they are effectively entirely resistant to amputation, to a phase in which they are extremely sensitive to the presence of the limb; if they get through this, they enter a third phase in which they become progressively less so (Prestige, M. C. [1967]. *J. Embryol. exp. Morph.*, **17**, 453). The cell that gets through the phase of extreme dependence on the limb and makes contact with the periphery has a tremendous advantage over all the other cells that haven't got through. This is a selective mechanism for ensuring that the surviving motor neurons have an efficient and stable connexion with the periphery, at the cost of a considerable cell redundancy.

Eccles: This establishes that the cells that are going to degenerate at amputation have already established connexions with the periphery, because otherwise how would they know to degenerate when the limb was amputated?

Prestige: I don't think that is necessary. Information could be passed from one cell body to another, by intercellular connexions.

Eccles: And this happens on that limb but not on the contralateral limb?

Prestige: Yes.

Kollros: If a piece of potential lumbar sacral cord from the frog embryo is grafted to the back of a very young larva in complete isolation from the limb, the lateral motor column will nonetheless differentiate. This has not been done in a hypophysectomized tadpole. If it had I would suspect that the lateral motor column would remain small but with very many cells indefinitely.

Hamburger: In the chick embryo if the leg bud or wing bud is extirpated at $2\frac{1}{2}$ days, normal differentiation and fibre outgrowth of all motor nerves occurs in the absence of the bud. The full complement of all the motor neurons (20,000 per limb) is established. The outgrowing fibres form a neuroma. After all neurons have been established and have reached a certain stage of differentiation, every one of them degenerates at between five and eight days although the fibres have never been cut or established any connexion. They respond to some kind of environmental deficiency which leads to a very rapid and very dramatic breakdown of the perikaryon within a few days. This happens only at these early stages—these are very young motor neurons. This suggests that environmental factors other than neuromuscular connexions are necessary to maintain a young embryonic motor neuroblast.

Székely: With Dr. Straznicky we repeated these experiments of removing the wing bud and counting the motor neurons in the ventral horn. We did not publish this work because we got exactly the same results: degeneration of motor neurons started around the fifth day and it was completed by the ninth day of incubation. This time coincides with that when the establishment of neuromuscular connexions begins. This suggests that motor neurons can go on to a certain stage of differentiation without any connexion to the periphery, but if the connexion is not established they die. Dr. Kollros's hypophysectomy experiment probably slows down the development and extends the time span during which the motor neurons can exist without any connexion, but after a certain stage they certainly die.

Hamburger: Your explanation is of course possible, but I should like to suggest an alternative, which however is unproved. My alternative is to relate this breakdown to the axoplasmic flow; obviously the spinning out of nerve fibres requires an enormous amount of protein synthesis in the perikaryon, corresponding to the axoplasmic flow in the adult fibre. It is conceivable that preventing the fibre from being spun out upsets the synthetic machinery in the perikaryon to such an extent that it breaks down, i.e. the axoplasm produced cannot flow out and cannot dissipate at the periphery, and very dense neuromas are formed. Perhaps this creates a pathological condition to which the perikaryon succumbs. It is not necessarily a neuromuscular relationship that is required for survival. It is possible that purely mechanical interference with axoplasmic flow has the same pathological effect.

Székely: The effect on nerve cell degeneration remains the same when one extirpates the wing bud on the fifth, sixth or seventh day.

Eccles: Dr. Hughes made a point about the possibility of the fibres having peripheral branches. They are not necessarily connected to anything, but in the amputation these cells register the fact that their axons are being cut and they all degenerate, regardless of whether they are going to be in the surviving or the non-surviving population.

Szentágothai: The surviving ones have their axons.

Eccles: Then the non-surviving ones also have their axons.

Kollros: Only 40 per cent of them at the most can possibly get their axons into the limb if the nerve fibre counts are correct.

Eccles: How are the nerve fibre counts done, and would the very fine non-medullated fibres be detected?

Kollros: The counts are done by silver staining, and Van Stone is pretty good at detecting those fibres.

Levi-Montalcini: In connexion with the problem of nerve cell degeneration during ontogenesis, I should like to call attention to the mass cell degeneration which takes place in many sectors of the developing nervous system (Hamburger, V., and Levi-Montalcini, R. [1950]. *Genetic Neurology*, pp. 128–160, ed. Weiss, P. Chicago: University of Chicago Press). One wonders about the cause of the sudden death of thousands of nerve cells which have already acquired distinct marks of differentiation such as the outgrowth of the axon.

Hughes: A wide spectrum of causes and relations are concerned in degeneration. Alfred Glücksmann (1951. *Biol. Rev.*, **26**, 59–86) found evidence of some basic distinctions between morphogenetic and histogenetic degeneration.

Gaze: Dr. Glücksmann's observations on the eye (1965. *Archs. Biol., Liège*, **76**, 419–437) are not confined to motor cells or internuncial cells. This phenomenon occurs also in the retinal cells during development of the frog's eye, where there are two or three waves of degeneration; virtually the whole of the central portion of the retina degenerates, to be replaced by cells coming in from the periphery. It may well be that this occurs also when degeneration takes place in the adult newt retina. Obviously it is a fairly general phenomenon.

Walton: Are the cells which disappear uniformly distributed throughout the ventral horns, or are they morphologically located in any particular part of the ventral horn? Secondly, are the nerve fibres at this stage myelinated and if so is there any correlation with fibre diameter at different stages at the time when the anterior horn cells are disappearing? It might be useful to carry out some work on osmic-fixed material, to see whether there is anything specific about the diameter of the fibres in the peripheral nerve which disappear at the same time as the anterior horn cells are disappearing.

Hughes: The peak of degeneration, which is so sharp in *Xenopus*, also follows the cranial-caudal sequence—it begins in front and works backwards. The fibres that we count are all enveloped in Schwann protoplasm but further differentiation hasn't yet occurred. In any case I think the whole thing waits on electron microscopy to see what is the spectrum of embryonic fibre diameters.

Drachman: We have been skirting the problem of what it is that the periphery supplies which prevents the anterior horn cells from dying or degenerating. It is clear that retrograde degeneration of anterior horn cells can occur after loss of the peripheral field even when there has unequivocally been no trauma to the axons. What is the contribution of the periphery that is necessary for maintenance of the anterior horn cells?

Hughes: I think there are two factors: firstly, the cell's reaction to axotomy and, secondly, the cutting off of something from the periphery.

Eccles: If Professor Young were here he would be able to talk about the relationship of the double-dependence story of nerves and muscles.

Drachman: But what is the language?

Eccles: Chemistry!

Singer: Does rapid death of cells of the anterior column also occur in the urodele amphibian?

Hughes: The urodele amphibian hasn't got a ventral horn, although it has motor horn cells and there are also commissural cells.

Singer: In connexion with the death of motor horn cells during development described by Dr. Hughes, some of our quantitative studies on the peripheral nerve supply of the limbs of various vertebrates may be of some interest (Singer, M., Rzehak, K., and Maier, C. S. [1967]. *J. exp. Zool.*, **166**, 89-98). The total number of fibres per unit cross-sectional area (100 μ^2) of the upper limb is approximately 25 in the newt, *Triturus*, but less than three in *Xenopus*, and it is similarly low in the mouse. Thus, the newt's limb is much more highly saturated with nerve fibres. However, the fibres themselves are much larger in *Xenopus*, the average axonal diameter being about 7·2 μ compared to 2·4 for the newt. The mouse also has small fibres. If instead of comparing number of fibres per unit cross-sectional area we now compare amount of axoplasm by multiplying number of fibres per unit area by the cross-sectional area of the average axonal diameter, we observe that the newt and *Xenopus* limbs contain similar amounts of neuroplasm. In this sense the limbs of the two animals are similarly innervated. Although *Xenopus* loses a lot of its cells during development, it makes up for the loss by an increase in fibre size. The newt probably loses much fewer neurons during development but has smaller fibres. Initial studies of the mouse limb suggest that in addition to fewer fibres there is much less axoplasm. Apparently the trend in evolution of the peripheral nerves was reduction in fibre number and amount of neuroplasm.

Eccles: This can of course all be seen when motor regeneration occurs. J. Z. Young (1962. *Physiol. Rev.*, **22**, 318) has shown that there are far more branches from the central stump than there were nerve fibres originally, and down each of the bands of Bünger several fine fibres grow out. Eventually in the process of full recovery one of them grows to full size and the others atrophy. There is a sorting-out, as it were, going on there too.

Singer: We have suggested that two evolutionary factors operated in establishing the extent of peripheral innervation. One was decrease in fibre number as better central mechanisms were elaborated. The second was maintenance of trophic nervous control of peripheral structures, which requires the delivery of an adequate volume of axoplasm. Both *Xenopus* and *Triturus* can regenerate their limbs, and regeneration of a limb requires an adequate threshold contribution of trophic substance from the nerves. In *Xenopus* the volume of the delivery is great enough because the axons are voluminous; in *Triturus* the delivery is satisfied by numerous, albeit less voluminous, fibres. In non-regenerating vertebrates (other frogs, reptiles and mammals) the fibres are neither abundant nor adequately voluminous.

Eccles: So far as I know the diameters of motor nerve fibres have never been correlated with the amounts of muscle innervated. There are some very important biological interactions of this kind that are concerned with how much muscle a nerve fibre can innervate and conversely how many innervated muscle fibres are necessary for the survival of the motor nerve fibres.

Gutmann: During development and regeneration there are always more axons formed than are retained. Also the degeneration of blastema in the brain during development is a normal occurrence and cell death apparently occurs as a normal developmental event (see Glücksmann, A. [1951]. *Biol. Rev.*, **26**, 59). When mitosis occurs in one blastema, the other connected with it may stop its mitotic activity. There are programmed interactions and one is faced with the concept of coding of time sequences in the development of the neurons and of muscles. "Programmed cell death" of muscles in insects during development (Lockshin, R. A. and Williams, C. M. [1965]. *J. Insect Physiol.*, **11**, 123) is another example.

Kollros: Dr. Drachman asked what kind of information the nervous system gets from the limb. Part of the evidence we have is that when the limb is removed in stages when lateral motor column cell counts are still large, more cells remain for a considerable time beyond their expected period of disappearance than is the case when the limb is present; so the question has to be rephrased—it is not what information the nervous system gets when the limb is present, but what information it gets when the limb is absent.

Eccles: Are you sure that when you amputate the limb you don't leave the nerve stump behind? What happens to that stump? Does it perhaps not regenerate to other environmental muscles in the body wall, for example.

Kollros: If it is a hypophysectomized animal it will remain as an early stage tadpole indefinitely, and the number of nerve cells present in the cord remains high indefinitely. But if we make the animal metamorphose then many of the motor cells die, which suggests that, if they do have a different terminal, the changed milieu of the added hormone is no longer an appropriate medium for maintaining them.

Drachman: Dr. Kollros's experiments show that motor neurons which have been exposed to the presence of a developing limb either develop connexions and survive, or fail to establish connexions and degenerate. On the other hand, "naïve" motor neurons, which have never been exposed to the influence of the limb, do not degenerate. Dr. Kollros has prolonged the innocence of some motor neurons either by transplantation of the spinal cord to a region remote from the limbs, or by hypophysectomy which prevents the development of limbs while leaving the cord in place. A possible explanation of these phenomena is that the limb's influence may be to commit the motor neurons to a new phase of development, such that they are obliged to innervate muscle or else will degenerate.

Hník: I was rather intrigued by what Dr. Prestige said about the three-phase course of resistance of the neurons to amputation. Is there any morphological correlation of these stages? Are the neurons, when they are highly resistant, in the neuroblast stage?

Prestige: Yes, the neuroblasts are the phase one cells. With reference to the point about whether the limb bud makes the phase one cells commit themselves to become phase two cells, can I argue with Dr. Kollros about the effect of isolated cords? There are reports in the literature (May, R. M. [1930]. *Bull. Biol. Fr. Belg.*, **64**, 355–387; Perri, T. [1956]. *Atti Accad. naz. Lincei Rc.*, **20**, 666–670; Hughes, A., and Tschumi, P. A. [1960]. *Jl R. microsc. Soc.*, **79**, 155–164) of isolated cords which have never been in contact with the limb but which maintain their ventral horn cells up to this phase one stage—the neuroblast stage—and then disappear. Those cells die or disappear without having had any trauma, as Professor Hamburger suggests; so these cells are definitely dying because of their failure to connect to the limb; and they are maturing from phase one to phase two independently of the limb, so the limb is not instructing the cells to commit themselves.

BIOLOGICAL ASPECTS OF THE NERVE GROWTH FACTOR

Rita Levi-Montalcini and Pietro U. Angeletti

Department of Biology, Washington University, St. Louis, Missouri

The biochemical properties of a protein and the growth response it elicits from sensory and sympathetic nerve cells have been the object of extensive studies. The results have been reported in detail (Cohen, 1960; Levi-Montalcini, 1952, 1964b; Levi-Montalcini and Angeletti, 1961; Levi-Montalcini and Booker, 1960) and will be only briefly summarized here.

The chance discovery that fragments of some mouse sarcomas implanted into the body wall of three-day chick embryos enhance growth and differentiative processes of sensory and sympathetic ganglia of the host (Bueker, 1948; Levi-Montalcini and Hamburger, 1951) was the starting point of an investigation which was to result in the isolation and partial characterization of the active agent released by the tumour (Cohen, Levi-Montalcini and Hamburger, 1954). The same agent was then discovered in much larger amount in two other biological sources, snake venom and mouse submaxillary salivary glands (Cohen and Levi-Montalcini, 1956; Levi-Montalcini and Cohen, 1960). The remarkable similarities in the biochemical properties and biological effects elicited by the protein isolated from these three sources seem to indicate that in all instances we are dealing with the same or with closely related proteins normally present in animal organisms and endowed with a potent and specific growth effect on the sensory and sympathetic nerve cells. To this protein, or proteins, we gave the name of Nerve Growth Factor (NGF) (Levi-Montalcini and Cohen, 1960).

Here we shall consider only the NGF isolated from mouse salivary glands. This proved in fact to be the richest source of this agent and for this reason it became the one used for the extraction of the NGF since its discovery in 1958 (Cohen, 1958). The similarities and differences between the NGF isolated from mouse sarcomas, snake venom and mouse salivary glands are reported in previous publications and will not be considered here. Suffice it to mention that this factor is present in snake venom and in mouse salivary glands at concentrations respectively 1,000 and 10,000 times higher than in mouse sarcomas (Cohen, 1958).

The active fraction isolated by Cohen from the mouse salivary glands is heat-labile, non-dialysable, destroyed by proteolytic enzymes, unstable to acid pH, but resistant to alkaline pH. It is endowed with antigenic properties and when injected with Freund's adjuvant induces the production of specific antibodies (Cohen, 1960). Its molecular weight, as determined with the analytical ultracentrifuge, is of the order of 44,000.

The analysis of the response of the target cells—the sensory and sympathetic nerve cells—to the NGF was performed in parallel series of investigations in the developing chick embryo, in newborn and adult mice and on sensory and sympathetic ganglia of different species, dissected out from chick and mammalian embryos and cultured *in vitro* in semi-solid and liquid media. For convenience, the chick embryo has been used for most of the morphological and biochemical studies performed *in vitro*, while the mouse has proved to be the best object on which to test the NGF in the newborn and adult animal.

When injected into the chick embryo, the NGF calls forth hypertrophic and hyperplastic responses from sensory and sympathetic ganglia. The excessive production of sensory and sympathetic nerve fibres by the enlarged ganglia results, in turn, in the hyper-innervation of cutaneous somatic fields and of most of the embryonic viscera (Levi-Montalcini, 1952). In the newborn animal (mouse) the growth effect is restricted to the sympathetic ganglia. An over-all enlargement of these ganglia up to ten to 12 times their normal size obtains in optimum cases in a ten-day period of treatment (Levi-Montalcini, 1966). The increase in volume of the sympathetic ganglia is due to an increase in nerve cell number and an increase in the size of individual neurons. Only the latter effect is present in adult mice injected with the specific growth factor.

The NGF was assayed *in vitro* on sensory and sympathetic ganglia dissected out from seven- to nine-day chick embryos and cultured in a semi-solid medium consisting of chicken plasma and Eagle's amino acid solution. The NGF added to this medium at a concentration of 10^{-8} g./ml. elicits the formation of a dense fibrillar halo in a six- to ten-hour period. The response to higher or lower concentrations of the NGF *in vitro* will be considered in detail in a later section (p. 133).

The metabolic effects of the growth factor were studied on sensory and sympathetic ganglia and on dissociated nerve cells cultured in liquid media for short periods ranging from one to eight hours. Labelled precursors were added to the culture media and their rate of incorporation was compared in control and NGF-treated ganglia or dissociated nerve cells. These experiments showed a marked enhancement of oxidative and synthetic processes in NGF-treated nerve cells. The increased rate of RNA synthesis precedes in time all other observed metabolic events.

The mode of action of the NGF was further analysed by studying the effects of specific inhibitors on the changes induced by the growth factor. Actinomycin D, which inhibits DNA-dependent RNA synthesis, interferes with the action of the NGF, suppressing its effects (stimulation of protein and lipid synthesis and increased glucose utilization) at all levels. These results suggested that the action of the NGF might be mediated through DNA-dependent RNA synthesis and raised the possibility that the growth factor acts on genetic transcription. Since the NGF is normally present in trace amounts in the responsive nerve cells and has proved to be essential to their survival (Levi-Montalcini and Angeletti, 1963), the suggestion was made that it might be part of a basic growth control mechanism of these cells (Levi-Montalcini and Angeletti, 1965).

In this condensed report on the morphological and metabolic effects of the NGF, we must omit other aspects of the phenomenon which have been under close scrutiny ever since large quantities of the NGF were discovered in the mouse salivary glands and the purified growth factor became available for more intensive and extensive studies on its nature, its source of production in the organism, and its mechanism of action. In this paper we shall consider only two of the many questions raised by the discovery of the nerve growth factor:

(a) What is the role of the mouse submaxillary gland in the production of the nerve growth factor?

(b) Does the NGF stimulate the synthesis of all cellular proteins or does it activate the production of a particular protein regardless of its precise role in the activating process? This question was approached from a biochemical, structural and ultrastructural viewpoint. Only the two latter points will be considered here.

ROLE OF MOUSE SUBMAXILLARY SALIVARY GLANDS IN THE PRODUCTION OF THE NERVE GROWTH FACTOR (NGF)

Since the discovery that mouse submaxillary salivary glands harbour a large amount of a protein endowed with nerve-growth-promoting activity, the question was raised of the role of these glands in its production. The NGF isolated from the mouse salivary glands was first identified in a protein with a molecular weight of 44,000 (Cohen, 1960). There is, however, clear evidence that the NGF exists in several molecular forms, possibly aggregates of a basic subunit. In fact, when a crude extract of the salivary glands is fractionated through a Sephadex column, the biological activity is found to be associated with fractions of high molecular weight of over 120,000 (Salvi, Angeletti and Frati, 1965).

Two alternative hypotheses were suggested to account for the high content of the NGF in salivary glands of the adult male mouse: (a) the NGF is cleared from the blood stream and concentrated in the tubular portion of the gland, (b) the NGF is produced *in situ*. Evidence to be presented in this paper favours the latter alternative.

Submaxillary salivary glands from adult male mice were dissected out and quickly sliced with a tissue microtome. The slices were weighed, placed in a Warburg flask containing 2 ml. of Eagle's basal medium supplemented with ^{14}C-labelled L-leucine and ^{14}C-labelled L-threonine (5 μc/ml.), and incubated for one or two hours. The slices were then quickly washed in several changes of a cold medium and homogenized in physiological solution. The homogenate was centrifuged at 10,000 rev./min. One portion of the clear supernatant was immediately precipitated with 5 per cent trichloroacetic acid and the precipitate was washed several times as in the method of Flexner and co-workers (1962). The final precipitate was dissolved in formamide and counted in a liquid scintillation system. Another portion (1 ml.) was added to the NGF with 2 ml. of antiserum (AS) and incubated for one hour at room temperature and 12 hours in a cold room. The antiserum to the most purified NGF, prepared by Cohen's (1960) method, had a titre of 1 : 1,000 when assayed *in vitro*. Before use, the antiserum was absorbed with fresh mouse liver tissue and with fresh salivary glands from prepubertal animals which still do not contain the NGF. The NGF-AS precipitate was washed several times with physiological solution and then the antigen–antibody mixture was dissociated in alkaline solution (Cohen, 1960).

Samples of this solution were tested in tissue culture for NGF activity. The remaining solution was precipitated with trichloroacetic acid, washed again as in the method of Flexner and co-workers (1962), finally dissociated in formamide and the radioactivity counted in a liquid scintillation system. The results shown in Table I indicate that under these experimental conditions, the slices of the salivary glands incorporate amino acids into protein and the incorporation rate is linear with time. A small percentage of the total radioactivity is found in the AS precipitate which yields NGF activity.

In a second series of experiments, 20 adult male mice were operated on under anaesthesia and the two lobes of the salivary gland were carefully separated from each other and exposed. Radioactive amino acids were injected into one of the two lobes with a microsyringe, while physiological solution in the same volume was injected into the other lobe. The animals were killed after two hours, the two lobes were separately homogenized, and experimental and control lobes were pooled in two groups and then

Table I

INCORPORATION OF RADIOACTIVE AMINO ACIDS IN TOTAL SOLUBLE PROTEIN AND IN THE NGF-AS PRECIPITATE IN TISSUE SLICES OF MOUSE SUBMAXILLARY SALIVARY GLANDS

Incubation time	Counts/min. per ml. total soluble protein	Counts/min., precipitated NGF-AS from 1 ml. homogenate
1 hr.	2,720 ± 320	115 ± 12
2 hr.	4,300 ± 640	250 ± 21

Three slices (60–80 mg. fresh weight) incubated in 2 ml. of Eagle's basal medium with added ^{14}C-labelled L-leucine and ^{14}C-labelled L-threonine, 2·5 μC of each/ml. One ml. of homogenate (5 mg. protein) reacted with 2 ml. of NGF-AS. Each value is the mean of three experiments.

centrifuged. The clear supernatants were fractionated through Sephadex G-100 and protein radioactivity and NGF activity were determined in the trichloroacetic acid precipitates from samples of each fraction. The protein labelling, as shown in Fig. 1, is about ten times higher in the lobe directly injected with radioactive precursors (Fig. 2). The radioactivity of the

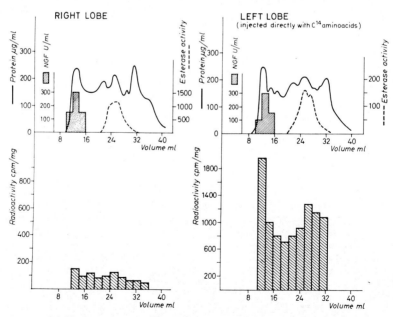

FIG. 1. Gel filtration patterns of proteins extracted from the right and left lobes of the mouse submaxillary gland. Labelled amino acids were injected directly into the left lobe in animals under anaesthesia (see text).

opposite lobe is of the same order as that found in several other organs of the same animals. The specific activity varied in the several fractions, as shown in Fig. 1, thus indicating the presence of proteins with slower or faster turnover. The NGF is localized in the earlier fractions, which were among the more heavily labelled. The fractions containing the NGF activity were then pooled, and processed for further purification of the NGF.

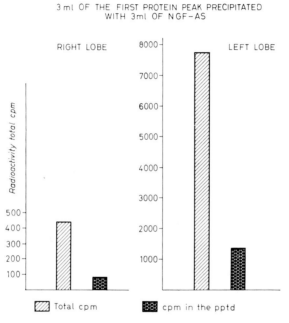

FIG. 2. Radioactivity in total soluble protein and in NGF-AS precipitate in the right and left lobe of the mouse submaxillary gland. Labelled amino acids injected into the left lobe.

The purification procedure consisted of two-step precipitation with ammonium sulphate, first at 50 per cent and then at 75 per cent. The precipitate obtained at 75 per cent was dissolved in a small volume of water and dialysed against several changes of tris buffer, 5 mM, pH 7·2, and finally against phosphate buffer 0·05 M, pH 6·5; it was then applied to a DEAE-Sephadex column, pH 6·8. A gradient salt elution from 0 to 0·3 M-sodium chloride solution was then used. Under these conditions the NGF weakly binds to the column and emerges in a sharp peak soon after the gradient starts. The fractions with the highest NGF activity were then pooled, concentrated and again put on a Sephadex G-100 column. From this column, the NGF emerges in one major peak well separated from minor contaminants of lower and higher molecular weight. Samples from this fraction with a specific biological activity equal to the most purified NGF were precipitated with antiserum and the precipitate was processed as described above for counting the radioactivity. The results, shown in Table II, indicate the differential labelling of the NGF from the two lobes.

The above experiments give additional evidence for the production *in situ* of the NGF present in the salivary glands, since one would expect the

Table II

INCORPORATION OF RADIOACTIVE LEUCINE AND THREONINE INTO PROTEIN FROM THE TWO LOBES OF THE MOUSE SUBMAXILLARY SALIVARY GLAND (20 MALES)

Protein fraction	Counts/min. per mg.		NGF activity (μg./ml.)
	Left lobe	Right lobe	
Total protein	1,320	125	2
First peak from Sephadex G-100	1,850	140	0·5
75 per cent ammonium sulphate fraction	2,100	165	0·25
DEAE Sephadex fraction	2,250	158	0·08
Second Sephadex G-100 fraction	2,180	162	0·04
NGF antiserum precipitate (100 μg. NGF fraction + 3 ml. antiserum)	250★	18	

The left lobe injected directly with radioactive precursors. Labelling for 2 hr. The NGF was purified as described in the text and then precipitated with antibodies. The NGF activity was assayed in tissue culture and expressed by the minimum protein concentration required to give a 3+ response.

★ Total counts/min. recovered in the precipitate.

NGF of the two lobes to have the same, or nearly the same, specific radioactivity if this protein were synthesized not in the salivary glands but elsewhere, and then transported to the salivary glands and stored there.

STRUCTURAL AND ULTRASTRUCTURAL ANALYSIS OF THE GROWTH RESPONSE
in vitro

The "halo effect"

Since our first study of the effects elicited by fragments of mouse sarcomas 180 and 37 on sensory and sympathetic ganglia of chick embryos cultured *in vitro* in a semi-solid medium, we have been impressed by the exceptionally dense and atypical growth pattern of nerve fibres from the entire surface of the explanted ganglia. Whereas under normal conditions only very few fibres grow out from the explant in the first day of culture and branch in an irregular fashion in the plasma, intermingling with fibroblasts and satellite cells, under the effect of adjacent fragments of neoplastic tissue the sprouting of nerve fibres undergoes drastic quantitative and qualitative changes. In a few hours the explant is surrounded by a halo of nerve fibres which grow radially away from the surface of the explant and produce a round or ellipsoidal fibrillar ring around the ganglion. The same and even more pronounced effects are elicited by the NGF isolated from snake venom and mouse salivary glands (Levi-Montalcini, 1958). The production of the fibrillar halo is in fact such a stereotyped and consistent performance from ganglia cultured in a semi-solid medium enriched with the NGF, that the "halo effect" was taken as unequivocal evidence of the presence of the growth factor in tissues or fluids added to the culture medium for the purpose of ascertaining whether they released the NGF, even in trace amounts.

In this way the NGF was detected in the blood, in extracts of a large number of organs and, in a somewhat larger quantity, in extracts of sympathetic ganglia of birds and mammals, man included (Levi-Montalcini and Angeletti, 1960). Since the dimensions of the halo and the density of the nerve fibres differ in relation to the amount of the NGF added to the medium, it became necessary to establish a growth-response curve which would offer an admittedly empirical but useful criterion to grade the response. We defined the halo, as depicted in Figs. 3d, 4b, as the growth response elicited by one biological unit of the growth factor. This response obtains when the salivary NGF is added at a concentration of 0·01 μg./ml. A two- to fivefold increase in the NGF concentration results in the production of a very short and dense fibrillar halo. A further increase of ten to 100 times in the concentration of the growth factor results in the production of exceedingly short fibres or in the total lack of outgrowth of nerve fibres. This effect, examined in ganglia *in vivo* or stained and mounted *in toto*, suggested that the NGF elicited an inhibitory effect when present at high concentration in the medium.

Closer inspection of a large number of ganglia cultured for 12 to 24 hours in the presence of the NGF, fixed and stained with a specific silver technique or toluidine and sectioned at 10 μ, offered a different explanation of the lack of outgrowth of fibres from the surface of ganglia cultured in media with high NGF content. The results to be briefly reported here are based on the analysis of hundreds of cultures of ganglia dissected out from eight-day chick embryos and explanted *in vitro* in a semi-solid medium (chicken plasma, amino acid solution and the NGF solution to be tested). The NGF was added in twofold decreasing concentrations from 1 μg./ml.

to 0·0001 µg./ml. Several controls were run at the same time. Control and experimental ganglia were fixed 12 to 24 hours after the beginning of the incubation period.

The results of a typical experimental series are given in Fig. 3. Drawings

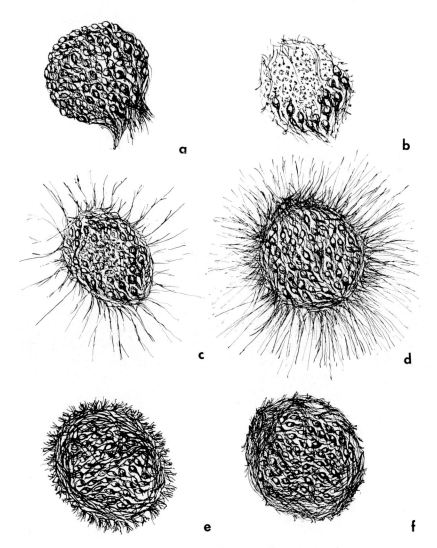

FIG. 3. Drawings from sections of ganglia dissected out from eight-day chick embryos and cultured *in vitro* in a semi-solid medium with progressively higher concentrations of NGF. All ganglia sectioned at 10 µ, fixed and stained with a silver technique. (a) control at zero time, (b) control fixed after 24 hours of culture. Note the disintegration of most nerve cells. For explanation of (c), (d), (e) and (f), see text.

BIOLOGICAL ASPECTS OF THE NGF 135

were preferred to microphotos for a more diagrammatic representation of the growth response.

As illustrated in Figs. 3c and 3d, the size and density of the fibrillar halo increases progressively as the concentration of the NGF in the medium is raised from 0·001 to 0·01 μg./ml. Fig. 3d shows the characteristic halo

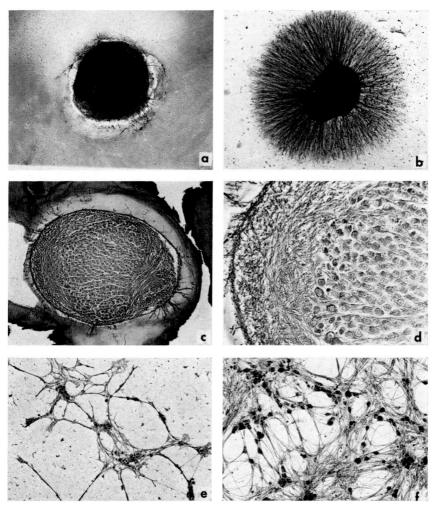

FIG. 4.(a) Whole mount of control eight-day sensory ganglion incubated 24 hours in a control semi-solid medium. (b) Whole mount of eight-day sensory ganglion incubated 24 hours in semi-solid medium to which the salivary NGF was added at a concentration of 0·01 μg./ml. (c) Section of an eight-day sensory ganglion incubated 24 hours in a medium to which salivary NGF was added at a concentration of 1 μg./ml. Note the dense fibrillar capsule and the absence of the "halo effect". (d) Detail of the fibrillar capsule of (c). (e) Dissociated sensory cells from eight-day ganglion incubated 24 hours in control liquid medium, and (f) in presence of NGF at a concentration of 0·1 μg./ml. Explanation in text.

which forms when the NGF content is 0·01 μg./ml. This is also defined as the effect of one biological unit of the growth factor. Figs. 3e and 3f show the effect of the same factor when it is added to the culture medium at concentrations of 0·1 and 1 μg./ml. respectively. The intermediate stages are omitted for the sake of brevity. It is apparent that the sharp decrease in the size of the fibrillar halo as represented in Fig. 3e, and its total absence in Fig. 3f, are compensated by a progressive increase in width and density of the fibrillar capsule which encircles the explant.

One further point deserves comment. The fibrillar halo characteristic of the effect of one biological unit of the NGF (Fig. 4b) appears *in vivo* and also after specific silver staining as an impressive structure of considerable body and texture. In reality it is built of very tenuous and thin fibres which stand out conspicuously *in vivo*, owing to their high optical refraction, and even more after staining, because of their high affinity for silver salts. The capsule which encircles the explant and is not apparent in the *in vivo* inspection or in ganglia stained and mounted *in toto*, is instead built of thick nerve fibres. Since this structure greatly increases in thickness and width at progressively higher NGF concentration, it follows that the production of neurofibrillar material keeps pace with the amount of growth factor present in the medium. Although quantitative studies should be made to ascertain this point, we may state that there is no indication of an inhibitory effect of the NGF on the outgrowth of nerve fibres when it is added to the culture medium at high concentrations. These results are consistent with studies made *in vivo* which showed that nerve cells receptive to the NGF have what seems to be an almost unlimited capacity to synthesize new neurofibrils under the impact of the growth factor. In the experiments *in vitro*, the circular rather than radial pattern of nerve fibre outgrowth in the presence of high NGF content remains to be explained, but this is rather irrelevant to the main point, namely the growth-response curve to progressively higher concentrations of the NGF.

Nerve fibre outgrowth from dissociated nerve cells in vitro

The technique for dissociating and culturing sensory and sympathetic nerve cells in liquid media was reported by Levi-Montalcini and Angeletti (1963). Here we shall only briefly mention the outcome of these experiments, which have a bearing on the problem of neurofibre production at different concentrations of the NGF. In its absence nerve cells deteriorate and die within the first 24 hours of culture. They survive and grow vigorously when the growth factor is added at concentrations ranging from 10^{-8} to 10^{-5} in the culture medium. At variance with the results reported

BIOLOGICAL ASPECTS OF THE NGF 137

in the previous section, no clear-cut differences were apparent in nerve fibre production at concentrations varying from 10^{-7} to 10^{-5}. In all instances the dissociated nerve cells produce a very large number of nerve fibres. One gets the impression that under these conditions the cells respond to their maximal capacity and no more nerve fibres are produced, perhaps

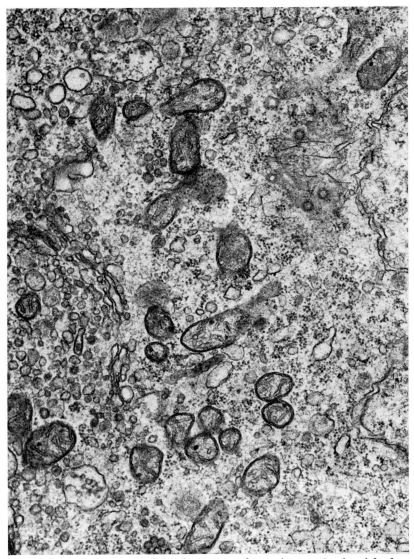

FIG. 5. Electron micrograph of cytoplasmic area of control neuron incubated for four hours in control semi-solid medium. Ribonucleoprotein particles arranged in rosettes. Nuclear envelope appears at right.

also due to the fact that all available space around the cells is filled with nerves. Fig. 4f illustrates the production of a dense fibrillar net of fibres after three days of culture in a medium where the NGF was present at the concentration of 10^{-5}. Fig. 4e shows a control culture. The results of both groups of experiments *in vitro* thus bring to the fore what seems to us the most outstanding and unique property of the NGF, namely the stimulation of the production of neurofibrillar material. This aspect of the phenomenon will be considered again in the discussion.

The NGF effect at the subcellular level

A detailed study of the fine structure of sensory nerve cells cultured *in vitro* for four to 12 hours, in the presence or absence of the NGF, is reported elsewhere (Levi-Montalcini et al., 1967). Here we shall only briefly consider the most relevant findings of this investigation.

Sensory ganglia were dissected out from eight-day chick embryos and explanted in a semi-solid medium in the same way as described in the previous section. The NGF was present in the culture medium at concentrations ten to 100 times higher than that which results in the halo effect. In other words, ten or 100 biological units of the growth factor were present in the culture medium. Control ganglia were fixed at zero time and after two and four hours of culture (Fig. 5). Experimental ganglia were fixed after two, four and 12 hours of culture (Fig. 6). The analysis of control ganglia after 12 hours *in vitro* was omitted, since histological studies of these ganglia already showed widespread signs of cell deterioration eight hours after the beginning of the culture period. At the end of incubation, control and experimental ganglia were very gently removed from the plasma clot and immediately fixed in cold glutaraldehyde. They were post-fixed for two hours in 2 per cent osmium tetroxide, dehydrated in ethanol solutions for five minutes each, placed in toluene for one hour and embedded in Epon. These sections were cut in an MT-1 microtome, stained with uranyl acetate and lead citrate, and examined in an RCA EMU-3H electron microscope.

The main feature of the growth response is the massive production of neurofibrillar material in the cell cytoplasm. This is already apparent two hours after the beginning of the incubation period and is in fact so extensive as to overshadow other less prominent ultrastructural changes, which will be only briefly mentioned here. The main observable changes in the cytoplasm of NGF-treated cells are a considerable increase in size of the Golgi complex and a disarray of the endoplasmic reticulum which, at variance with controls where it appears as regularly and evenly spaced parallel lamellae, consists of irregularly spaced lamellae with short side branches. At the outer surface of the lamellae are studded ribonucleoprotein particles which appear more numerous than in controls. Prominent changes are also apparent in the nucleus after a short exposure to NGF. The nucleo-

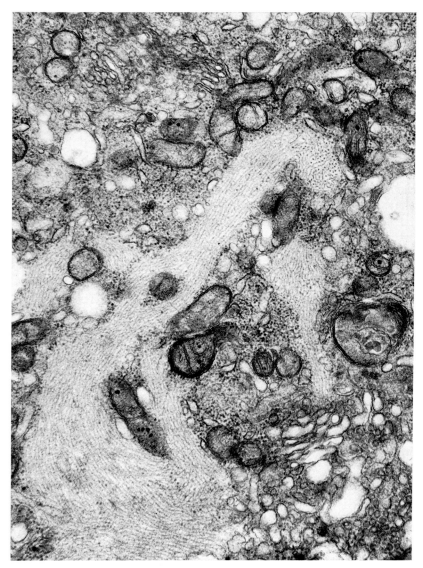

FIG. 6. Electron micrograph of cytoplasmic area of neuron cultured for 12 hours in a medium to which the NGF was added at a concentration of 0·1 µg./ml. The dilated Golgi complex is evident at the lower right part of the field. Large neurofilament masses fill part of the field.

plasm in these cells is more dense than in controls and aggregates of more electron-dense chromatin are irregularly dispersed throughout. The nuclear contour is regular and circular or oval-shaped in controls, while in experimental cells it shows numerous and long digitations. A possible change in the membrane pores is now under investigation.

DISCUSSION

In the first part of this paper we presented evidence showing that the NGF present in large quantity in the tubular portion of the male mouse salivary glands is the product of local synthesis. In previous articles two alternatives were submitted: the salivary NGF is produced elsewhere and then stored in the large tubules of the gland, or it originates *in situ* (Levi-Montalcini, 1964a). The first alternative seemed to us the more likely. While we now believe that we have decisive evidence in favour of the second alternative, we have no explanation to offer for such large-scale production of the NGF in the mouse salivary glands. We shall list here only the main arguments, presented in detail elsewhere (Levi-Montalcini and Angeletti, 1964; Levi-Montalcini, 1966), which oppose the concept of the salivary glands as endocrine organs releasing the NGF in hormonal fashion.

Removal of these glands in mice does not produce any detrimental effect on the sympathetic nerve cells, and yet we have decisive evidence that these cells depend for their growth and maintenance on this factor (Levi-Montalcini and Angeletti, 1963). Since the classic and most reliable test for identifying the origin of a given hormone is that hormonal deficiency effects follow the exclusion of the endocrine organ which manufactures the hormone, the only other explanation would be that the same factor is produced by other tissues or organs. We have positive evidence that this is indeed the case in our system (Levi-Montalcini and Angeletti, 1960; Levi-Montalcini, 1966). Against the hypothesis that the salivary NGF is produced with the precise aim of providing it to sensory and sympathetic nerve cells, is the finding that, of all mammals, only the mouse salivary glands harbour this agent; yet we have evidence that sensory and sympathetic nerve cells of other species, man included, are receptive to its growth-promoting activity (Levi-Montalcini and Angeletti, 1960). The question is raised of whether this protein might not also serve other purposes connected with the digestive function of this gland, or perhaps with its other main function, which is to produce and release toxic substances in the saliva. It is of interest in this connexion to remember that the same protein is also found at a very high concentration in snake venom.

In the second part of this paper we have analysed the growth response of embryonic sensory nerve cells to the NGF at the structural and ultrastructural level. Both sets of experiments indicate that the most dramatic effect elicited by the NGF is the massive production of neurofibrillar material. This effect is in fact so impressive as to overshadow all other effects. These and other experiments which gave evidence for a differential stimulation of cell proteins by the NGF (Gandini-Attardi, Calissano and Angeletti, 1967) suggest that the main function of the NGF is specifically to stimulate the production of a given class of proteins which, in turn, provide the building blocks for manufacturing the unique structure which is the neurofibril.

Finally, we should like to comment on what perhaps is the most challenging and still least understood aspect of the whole problem. The salivary glands have proved to be a source not only of the nerve growth factor but also of other equally potent and specific factors. Besides the NGF, another protein was isolated and purified from the salivary gland homogenate. This protein, discovered and characterized by Cohen, was found by him to exert a potent and specific growth-promoting activity on epidermal cells. It was named, after its effects, the epidermal growth factor (Cohen, 1962). Recently still another protein was isolated by us from the same glands. Upon further purification and characterization the protein was found to increase greatly the number of granulocytic cells in the blood of newborn and adult mice and also of all other mammals which were investigated. Preliminary experiments *in vitro* and *in vivo* seem to indicate a direct stimulatory action of this protein on bone marrow precursors of these cells. These findings, while providing additional evidence against the assumption that the salivary glands should be listed among endocrine organs, do not decrease but rather add to the complexity of the phenomenon under investigation. An intensive search for other growth-specific factors in this and other glands and tissues is now in progress in our laboratory. It is hoped in this way to obtain new information and perhaps eventually also the key to this still largely unexplored area of regulatory mechanisms of cell growth and differentiation.

SUMMARY

Evidence is presented that the nerve growth factor (NGF) markedly stimulates the growth of embryonic sensory and embryonic and mature sympathetic nerve cells. The *in vivo* and *in vitro* effects of this agent, which was identified in a protein particle, were investigated at morphological and metabolic levels. The results of these investigations are briefly presented.

Recent data on the role played by the mouse submaxillary gland in the production of NGF show that this protein is actually synthesized in the gland (striated ducts) and is largely excreted with saliva; its real physiological role in the gland remains to be explained. Observations on the ultrastructural changes brought about by the NGF on the receptive nerve cells *in vitro* are also included.

Acknowledgements

This investigation was supported by grants from the John A. Hartford Foundation and from the U.S. Air Force (AF-AFOSR-1182-67).

REFERENCES

BUEKER, E. D. (1948). *Anat. Rec.*, **102**, 369–390.
COHEN, S. (1958). In *Chemical Basis of Development*, pp. 665–677, ed. McElroy, W. D., and Glass, B. Baltimore: Johns Hopkins.
COHEN, S. (1960). *Proc. natn. Acad. Sci. U.S.A.*, **46**, 302–311.
COHEN, S. (1962). *J. biol. Chem.*, **237**, 1555–1562.
COHEN, S., and LEVI-MONTALCINI, R. (1956). *Proc. natn. Acad. Sci. U.S.A.*, **42**, 571–574.
COHEN, S., LEVI-MONTALCINI, R., and HAMBURGER, V. (1954). *Proc. natn. Acad. Sci. U.S.A.* **40**, 1014–1018.
FLEXNER, J. B., FLEXNER, L. B., STELLAR, E., DE LA HABA, G., and ROBERTS, R. B. (1962). *J. Neurochem.*, **9**, 595.
GANDINI-ATTARDI, D., CALISSANO, P., and ANGELETTI, P. (1967). *Brain Res., Amst.*, **6**, 367–370
LEVI-MONTALCINI, R. (1952). *Ann. N. Y. Acad. Sci.*, **55**, 330–343.
LEVI-MONTALCINI, R. (1958). In *Chemical Basis of Development*, pp. 646–664, ed. McElroy, W. D., and Glass, B. Baltimore: Johns Hopkins.
LEVI-MONTALCINI, R. (1964a). *Science*, **143**, 105–110.
LEVI-MONTALCINI, R. (1964b). *Ann. N. Y. Acad. Sci.*, **118**, 149–170.
LEVI-MONTALCINI, R. (1966). *Harvey Lect.*, **60**, 217–259.
LEVI-MONTALCINI, R., and ANGELETTI, P. (1960). In *Regional Neurochemistry*, pp. 362–377, ed. Kety, S., and Elkes, J. New York: Macmillan and Pergamon.
LEVI-MONTALCINI, R., and ANGELETTI, P. (1961). *Q. Rev. Biol.*, **36**, 99–108.
LEVI-MONTALCINI, R., and ANGELETTI, P. (1963). *Devl Biol.*, **7**, 653–659.
LEVI-MONTALCINI, R., and ANGELETTI, P. (1964). In *Salivary Glands and their Secretions*, pp. 129–141, ed. Sreebny, L. M., and Meyer, J. New York: Macmillan.
LEVI-MONTALCINI, R., and ANGELETTI, P. (1965). In *Organogenesis*, pp. 187–198, ed. DeHaan, R. L., and Ursprung, H. New York: Holt.
LEVI-MONTALCINI, R., and BOOKER, B. (1960). *Proc. natn. Acad. Sci. U.S.A.*, **46**, 384–391.
LEVI-MONTALCINI, R., CARAMIA, F., LUSE, S., and ANGELETTI, P. (1967). *Brain Res., Amst.*, in press.
LEVI-MONTALCINI, R., and COHEN, S. (1960). *Ann. N. Y. Acad. Sci.*, **85**, 324–341.
LEVI-MONTALCINI, R., and HAMBURGER, V. (1951). *J. exp. Zool.*, **116**, 321–362.
SALVI, M. L., ANGELETTI, P., and FRATI, L. (1965). *Farmaco*, **20**, 12–21.

DISCUSSION

Kerkut: In radioautography with the electron microscope Dr. B. Droz (1967. *J. Microsc.*, **6**, 201–228) can show incorporation of leucine into the neurofilaments. This technique might be used to show the effect of NGF on incorporation.

Levi-Montalcini: This may prove to be a very valuable technique also in our case. Since the NGF greatly increased the neurofibrillar material in the receptive nerve cells one would expect a markedly increased incorporation of labelled precursors in the treated nerve cells.

Eayrs: Is the NGF which you obtained from mouse tumour chromatographically identical with that which you obtained from the salivary glands?

Levi-Montalcini: The NGF is present only in trace amounts in the extract of mouse sarcomas 180 and 37 and for this reason it was not possible to extract enough NGF from these tumours and perform chromatographic studies. I cannot therefore answer your question. I wish to add that we have evidence that the NGFs isolated from snake venom and mouse salivary glands are immunologically and biochemically similar but not identical.

Eayrs: It is curious that dorsal root ganglia and autonomic ganglia seem particularly susceptible to this agent.

Is there some affinity between the NGF and all the derivatives of the neural crest? These can be found in a variety of other sites in the body, some of which are not strictly neural.

Levi-Montalcini: Only two neural crest derivatives, namely the sensory nerve cells of the cephalic and spinal root ganglia, are receptive to the action of the NGF. We have presented evidence that not even all sensory and sympathetic nerve cells respond to this growth factor (Levi-Montalcini, R. [1966]. *Harvey Lect.*, **60**, 217–259). Other neural crest derivatives, such as the parasympathetic nerve cells, pigment cells, satellite cells and cells of the adrenal medulla, are not receptive to the action of the NGF.

Gutmann: You said that there are two groups of ganglion cells, one responding, the other not. There are two groups of spinal ganglion cells in mammals—large ones and small ones—and there are also related differences in cholinesterase activity. Does this differentiation exist also in the ganglion cells of the chicken?

Levi-Montalcini: In the sensory ganglia of mouse embryos and other mammalian embryos there are cholinesterase-rich and cholinesterase-poor cells, as in the same ganglia in birds. They are not so clearly segregated in two distinct cell populations as in the chick embryo and we did not investigate the effect of the NGF on these two cell types but it is likely that it is the same as in the chick.

Gutmann: Are there also two populations according to size?

Levi-Montalcini: The size difference of nerve cells of these two populations is clearly apparent only during early embryonic life but not in later embryonic and post-embryonic stages. We do not know if any functional significance attaches to these morphological and biochemical differences between sensory nerve cells.

Szentágothai: How do the ganglion cells of placodal origin react?

Levi-Montalcini: The trigeminal ganglion consists of two nerve cell populations. One has its origin in one of the cephalic placodes, the other in the neural crest. The latter but not the former responds to the NGF. This is true also for other ganglia of placodal origin such as the facial, the petrosum and nodosum:

they are not receptive to the NGF. The lens is likewise not affected by this growth factor.

Szentágothai: Among the spinal ganglion cells only 10 to 20 per cent appear to have an axon. The number of cells is much larger than the number of fibres. Those responding might be those which have no axons.

Levi-Montalcini: No. We have definite evidence that nerve cells receptive to the NGF are provided with an axon. When mouse sarcomas were implanted in the chick embryo, the tumour was massively invaded by sensory and sympathetic nerve fibres. Both sensory and sympathetic nerve fibres were traced to their cells of origin in the spinal root ganglia and in the adjacent sympathetic ganglia.

Drachman: One of the most important questions regarding nerve growth factor is whether it plays a significant role in *normal* embryonic development. The observation that anti-NGF antibody impairs the development of the sympathetic nervous system *in vivo* has always seemed to me to be one of the most convincing pieces of evidence in favour of the factor's role in normal development. However, M. T. Sabatini, A. P. de Iraldi and E. De Robertis (1965. *Expl Neurol.*, **12**, 370–383) have shown that anti-NGF antiserum exerts a *direct* cytotoxic effect on sympathetic ganglion cells, which resembles other antigen-antibody reactions. These observations would not favour the view that the antiserum deprives the sympathetic cells of a naturally-occurring essential growth factor, but rather that it exerts a direct effect on some antigen within the cells themselves.

Levi-Montalcini: The experiment you mention by Sabatini, de Iraldi and De Robertis was performed on sympathetic ganglia cultured *in vitro* for two to twelve hours in a medium containing the antiserum to the NGF. The electron microscopic study of the treated ganglia showed severe damage to sympathetic nerve cells, suggestive of a cytotoxic effect.

Drachman: The alternative explanation is that there is a common antigen shared by NGF and by the cells of the sympathetic ganglia, since both are susceptible to specific antibody. If there is simply a common antigen, then the effects of antiserum cannot be used as evidence that NGF serves as a specific nerve-growth-promoting hormone in *normal* development.

Levi-Montalcini: I agree with you that the destructive effect of the antiserum does not necessarily imply that the NGF plays an essential role in the life of the receptive nerve cells. The suggestion that the NGF in fact plays this role came from a different set of experiments described in detail in a previous publication (Levi-Montalcini and Angeletti, 1963, *loc. cit.*). We reported that sensory and sympathetic nerve cells when dissociated and dispersed in the minimum of liquid medium die and distintegrate in a matter of ten to twelve hours. The same cells survive and grow vigorously for up to a month upon addition of the NGF to the culture medium at the very low concentration of 10^{-8} g./ml. We consider these results as indicative of the all-important role played by the NGF in the life of the receptive nerve cells.

DISCUSSION

Crain: You showed that the culture of an explant of sarcoma tissue near an explant of ganglion would produce an NGF halo. If the mouse salivary gland also produces this factor have you considered growing pairs of explants of salivary gland and ganglia in similar fashion?

Levi-Montalcini: We performed this experiment several times. The results were negative and we believe we have a satisfactory explanation for it. The salivary glands of adult mice release, among other things, a large quantity of proteolytic enzymes which digest the plasma clot in the proximity of the explant and also damage the cells in the adjacent ganglia. Furthermore the NGF is present in such a large amount in the salivary explant that an inhibitory rather than a growth effect results. The inhibitory effect of large doses of NGF is discussed in detail in my paper here.

Drachman: When we followed Dr. Cohen's scheme for purification of nerve growth factor we found that the 50 per cent ammonium sulphate-precipitated fraction produced inhibition of NGF-stimulated growth at dilutions of 10^{-4} to 10^{-5}. At 10^{-6} dilution the effect was lost.

Levi-Montalcini: Did you make cytological studies of the ganglia cultured in the presence of the two fractions of the salivary gland?

Drachman: No, only on the growth level.

Levi-Montalcini: A cytological study would have been useful to decide whether a growth effect is present in spite of the lack of the fibrillar halo (see my paper). If this "inhibitory fraction" is effective at different dilutions of the same fraction one should conclude, as you did, that it has an inhibitory or a toxic effect. This is, however, difficult to explain since in our experiments the whole extract of the mouse salivary gland, which also contains the 50 per cent precipitate, elicits a potent growth-promoting effect on sensory and sympathetic ganglia *in vitro*.

Eccles: This fascinating story worries me in one respect. The nerve growth factor is quite remarkable in the specificity of the target neurons. It apparently doesn't work on other parts at all. You have looked through such strange regions of the nervous system as exist, for example, in the hypothalamus.

Levi-Montalcini: The hypothalamus is not affected by the NGF, nor for that matter is any other part of the central nervous system.

Eccles: What about the retina?

Levi-Montalcini: The retina likewise does not respond to the NGF. The trigeminal ganglion consists of large cholinesterase-rich cells of placodal origin and of small cholinesterase-poor nerve cells. The latter but not the former are receptive to the NGF.

Eccles: Are the large cholinesterase-rich cells also in the dorsal root ganglion?

Levi-Montalcini: These cells of placodal origin in the trigeminal ganglion are structurally similar to the cholinesterase-rich cells of spinal root ganglia. In the latter ganglia, however, these cells originate from the neural crest. We know nothing of the functional significance of these two sensory nerve cell types.

Eccles: How do all the other parts of the nervous system grow if they have no NGF?

Levi-Montalcini: We believe that other specific growth factors for other nerve cell types could exist even if they have not been detected so far. This is one of the aspects of the problem which we are now investigating.

Eccles: The problem now remains, how do the rest of the cells in the nervous system get by? It surely is a major problem to discover NGF factors for them all, factors which may vary from the one you have. But you have no evidence whatsoever on this?

Levi-Montalcini: As I briefly mentioned above, we have not until now discovered any other specific nerve growth factor besides the one under discussion. However, another specific growth factor was discovered by Dr. S. Cohen in the extract of the mouse salivary glands. This factor, named from its action the epidermal growth factor (EGF), elicits a very potent and specific growth effect on epidermal cells (Cohen, 1962, *loc. cit.*). Another factor with a specific effect on granulocytic white cells is at present under investigation in our laboratory. These findings encourage our search for other specific growth factors and in particular for factors acting on nerve cells.

Szentágothai: Have you tried the effect of the growth factor on cultures of the central nervous system?

Levi-Montalcini: Yes, we tried it on fragments of different parts of the central nervous system, but the results were consistently negative.

Drachman: Some time ago we tested the effects of a number of growth-producing or neuroactive substances on ganglion cultures and found that none of them mimics the activity of NGF. The series included acetylcholine, Mecholyl, bradykinin, isopropylarterenol, serotonin, γ-aminobutyric acid, human growth hormone and thyroid hormone. Have you found any growth-promoting effect of any similar substances?

Levi-Montalcini: We tested a number of pharmacologically active substances and neurohormones, adding them at different concentrations to the culture media. In no case did we obtain the fibrillar halo as with the NGF, nor did we see any sign of a growth effect.

Eayrs: If we are to rely on a variety of NGFs, each matched to a specific cell, where are these to be manufactured?

Levi-Montalcini: Our working hypothesis is that specific growth factors might be manufactured everywhere in the organism and not in any particular gland as is the case for hormones. They are possibly metabolites released by cells of different types and utilized by other cells as growth factors. In the case of the NGF we have positive evidence not only that it is released by some neoplastic cells and is present in large quantity in the mouse salivary glands, but also that it is released by mesenchymal cells in active growth phase (Levi-Montalcini and Angeletti, 1960, *loc. cit.*).

Murray: Must growth factors for all kinds of nerve cells come from the salivary gland? Would it be worth while looking at some other glands?

Levi-Montalcini: This is in fact what we are doing at present.

Eccles: Is the NGF found in the salivary glands of all mammals?

Levi-Montalcini: No; ever since we found it in large quantity in snake venom and in mouse salivary glands, we have searched for it in the salivary glands of other species and in particular in mammals. The results were negative. Only recently have we found it in discrete quantity in the rat's salivary glands.

Eccles: In animals without the NGF in their salivary glands do the sympathetic ganglion cells and the dorsal root ganglion cells still respond to it?

Levi-Montalcini: In spite of the fact that the NGF is not detectable in the salivary glands of other species, sensory and sympathetic nerve cells of all species investigated, man included, are receptive to the action of the NGF.

Eccles: What is the site of production?

Levi-Montalcini: We believe that it might be produced by actively growing mesenchymal cells, as I mentioned above, but this is at present far from well substantiated.

Singer: What are these cultures like in relation to satellite cells? If the substance goes into the neuron by way of the satellite cell, that might account for the difference between the central versus the peripheral response, because in the periphery there is the Schwann-related satellite cell whereas in the centre there is another type of cell.

Levi-Montalcini: We believe that we have decisive evidence against this possibility. The NGF is in fact very effective not only on whole ganglia where satellite cells are in close contact with nerve cells, but also on dissociated nerve cells where these contacts between satellite and nerve cells are lost.

ACTION OF HEAVY WATER [D$_2$O] ON GROWTH AND DEVELOPMENT OF ISOLATED NERVOUS TISSUES

Margaret R. Murray and Helena H. Benitez

Departments of Anatomy and Surgery, College of Physicians and Surgeons, Columbia University, New York

Mice subjected to moderate degrees of hydrogen replacement by deuterium administered as D$_2$O in their drinking water have been reported to show behavioural disturbances. A review by Barbour (1937) recounts the early exploration of biological effects of D$_2$O on small mammals. Within the last decade a new method of preparation has greatly reduced isotope costs, and systematic investigations have been resumed in the higher organisms. In mice and rats (Thomson, 1963) overt signs of D$_2$O intoxication are not observed until about 15 per cent of the body water has been replaced; above this level the animals become hyperexcitable and hard to handle. In the range of 20 to 25 per cent D$_2$O, combativeness is very pronounced and convulsions may occur after even mild stimulation. Mice show increased pilomotor activity and exophthalmos, which Barbour and Herrmann (1938) interpreted as signs of sympathetic erethism, since these symptoms can be simulated by adrenaline (epinephrine) and are reversed by ergotoxin. When 30 per cent of the body fluids have been replaced the animals become comatose, and they die before the D$_2$O concentration in body water reaches 35 per cent. Mice can be maintained for several months with no apparent ill effects at 20 per cent saturation, though showing an elevated metabolic rate and body temperature.

Histological studies on deuterated animals have been scattered. Bachner, McKay and Rittenberg (1964) performed necropsies on mice killed after five days of exposure to 50 per cent D$_2$O; they found no gross or microscopic anatomical alterations, though the animals had exhibited "sympathicomimetic" behaviour. But in mice which had died spontaneously after drinking 75 per cent D$_2$O for six to ten days, these observers found gross and microscopic lesions in kidneys, testes and submandibular salivary glands; the nervous system was grossly normal. The animals had from 21 to 37 per cent D$_2$O in their body water at death. Katz and co-workers

(1962) report that incorporation of deuterium into the brain of mice is much slower than into kidney and other organs.

Since it has been shown by Levi-Montalcini (1966) and her co-workers that the submandibular gland is the repository of a variety of tissue-specific growth-promoting substances—notably the Nerve Growth Factor (NGF)—we supposed that a meaningful relationship might exist between the lesions of the salivary gland—especially of its ducts—and the "sympathicomimetic" action of D_2O *in vivo*. We therefore tested this hypothesis at the cellular level by exposing sympathetic ganglia, isolated in culture, directly to heavy water and separately to NGF (Murray and Benitez, 1967). Since the effects of these two compounds on ganglion tissues differed both quantitatively and qualitatively, it became apparent that our inquiry should be focused upon the action of deuterium, and probably on its cytological manifestations in the neuraxis as a whole.

METHODS AND MATERIALS

The Maximow double-coverslip assembly with reconstituted collagen as the immediate substrate was used throughout (Bornstein and Murray, 1958). Explants (about 1 or 2 mm.^{3}) were maintained as organized, developing cell populations during three weeks to three months as required, without transfer or other disturbance of the pattern of their cellular interrelationships, while being exposed continuously to D_2O incorporated in concentrations of from 5 to 33 per cent into their total feeding medium. Controls received equal portions of balanced saline solution (BSS), normal water or (in early experiments) NGF at 2 units/ml. All cultures were washed twice weekly in BSS, and their medium was renewed. The basic biological feeding formula consisted generally of equal parts of human placental serum, bovine serum ultrafiltrate or Eagle's medium, and saline extract of nine-day chick embryos; to this mixture was added glucose to reach a total concentration of 600 mg./100 ml. Minor variations in details of the formula were made, to comply with the requirements of different regions of the nervous system.

Superior cervical, stellate and sympathetic chain ganglia were explanted from embryonic chicks (14 to 16 days) and mice of 17 to 18 days gestation (about 500 cultures in all).

Dorsal root ganglia (sensory) were obtained from 11-day chick embryos, 17-day mouse foetuses and 18-day rat foetuses (256 cultures in all); some were maintained up to 76 days *in vitro*.

Hypothalamic explants came mainly from newborn mice and 17- to 18-day foetuses (125 cultures); 25 cultures from 12- and 17-day chick embryos.

The *cerebellum* from eight newborn mice was divided into four explants: 32 cultures were studied for 21 days *in vitro*.

Cerebral neocortex from three-day-old mice provided 125 cultures in all; some were maintained for as long as 73 days.

The above neural areas were explanted at the respective stages of development that proved most favourable in our hands for observations of their growth, differentiation and function *in vitro*.

One series of *submandibular* and *parotid glands* from newborn mice was explanted and cultured separately under the same conditions as the nervous tissue (32 cultures, maintained for 37 days).

The neural area most intensively investigated in these experiments is the autonomic nervous system. In addition to observations made almost daily with the light microscope, and frequent photography of living cultures, over 1,000 electron micrographs have been examined. For these, control and deuterated cultures were fixed either with 2 per cent osmium tetroxide in Veronal acetate buffer (pH 7·4), or with 3.5 per cent glutaraldehyde followed by osmication; the fixatives were prepared either with normal H_2O, or with a portion of the H_2O replaced by an amount of D_2O equivalent to that in the culture medium. After fixation, the cultures were dehydrated through an ethanol series and propylene oxide, embedded in Epon 812, thin-sectioned with a diamond knife, and stained with 50 per cent ethanolic uranyl acetate plus Reynolds' lead citrate.

OBSERVATIONS

Histologically different regions of the neuraxis vary in their responses to D_2O, as regards both behavioural details and optimum concentrations for the agent. Microscopical and ultrastructural aspects, and some functional attributes, of sympathetic ganglia in long-term organized culture will be described here in depth. Comparative observations on other areas will be recounted briefly.

Sympathetic ganglia

Controls. The developmental sequence for ganglia cultured in the routine biological media supplemented with a fraction of BSS or normal water can be summarized as follows:

At first, scattered bursts of fine neurites form a smooth outgrowing network after 48 hours; their interneuritic spaces are then slowly invaded by cells of *supporting-tissue* type, until by *nine to ten* days the explant has thinned sufficiently for the relatively immobile neurons to be discernible. These consist of groups of very small cells so closely packed inside capsules as to appear epithelial; at this time the round nucleus is surrounded by a thin rim of cytoplasm, and the whole cell may be 12 μ in diameter. Next, these neuroblasts enlarge, with predominant increase in cytoplasmic volume, and the capsule becomes less confining. As growth and maturation proceed, the young neurons become multipolar and discrete, and a quota of satellite cells associate themselves rather loosely with them. With increase in size to about 20 μ in diameter, the neurons perforce shift somewhat in relative positions but do not migrate far. This degree of maturity is reached by chick material at about 30 days and in mouse a few days earlier, probably because the murine neuroblasts are initially larger and less confined by capsules than their avian counterparts. Cultures may subsequently remain in this state, with slight, gradual increase in neuron size and dendritic complexity, for two, three or four months. Neurons of birds and mammals generally do not multiply, *in vivo* or *in vitro*, after having passed the early stages of normal maturation (Murray, 1965).

Experimental. D_2O in concentrations from 5 to 25 per cent produces an immediate acceleration of growth and maturation in both neurons and supporting cells. There is a perceptible concentration gradient in this response which peaks at 25 per cent. The isotope, when continuously supplied at these levels in the ambient medium, is well tolerated for as long as cultures have been carried (up to four months). However, ganglia explanted in 33 per cent D_2O put forth a phenomenally rapid and extensive outgrowth of neurites which are unaccompanied by supporting cells. The explant itself remains compact and in about two weeks becomes necrotic.

In deuterated sympathetic ganglia maintained at tolerable levels, outgrowing neurites are very long and branching as compared to the controls, although less numerous and bushy than in the characteristic NGF halo (Levi-Montalcini, 1964). Outgrowth of the deuterated supporting tissue (chiefly Schwann and satellite cells) is especially profuse and shows many mitotic figures. As soon as the neurons can be recognized in the explant (at about eight days *in vitro*) their somas are significantly enlarged, having a diameter of 12 to 18 μ. At the same time, capsules containing small, undifferentiated, epithelioid nerve cells in oval or circular clusters are emerging along the edge of the interganglionic portion of the chain—a phenomenon that does not characterize control cultures at this stage. After 25 to 30 days *in vitro*, deuterated neurons are three to four times the size of the controls; some neurons may have reached a giant size (about 70 μ in diameter) that is never seen in control circumstances. (At this time neurons of ganglia cultured in NGF are roughly twice the control size: Crain,

Benitez and Vatter, 1964). As the deuterium treatment is continued the neuron population increases markedly, so that after 60 days in culture it is estimated at two to three times that of control ganglia.

Actual cell counts are difficult, and rather undependable because of the thickness necessary to the ganglion explant—some six cell layers. However,

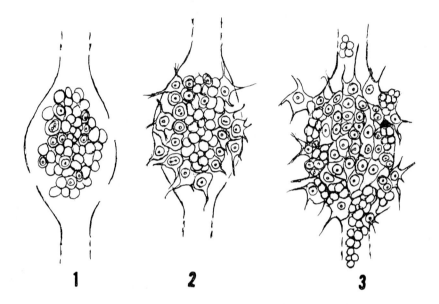

FIG. 1. Diagrammatic representation of stages in growth and maturation achieved by sympathetic ganglia *in vitro*.
   Stage 1 is reached by controls in ten days.
   Stage 2 is reached by controls in 20–30 days; by deuterated cultures in ten days.
   Stage 3 is reached by deuterated cultures in 30 days; BSS controls rarely reach this; H_2O controls sometimes may achieve this amount of growth and increase in number in 60–90 days.

Table I

COMPARATIVE RATES OF DEVELOPMENT (STAGES 1–3): SYMPATHETIC GANGLIA (FROM CHICK 14 TO 16 DAYS *in ovo*)

Medium modification	Days *in vitro*		
	8	21	54
BSS 25%	1·0	1·5	2·0
H_2O 25%	1·0	2·0	2·5
D_2O 25%	1·8	3·0	3·0 (×2)
5%	1·5	3·0	3·0 (×2)

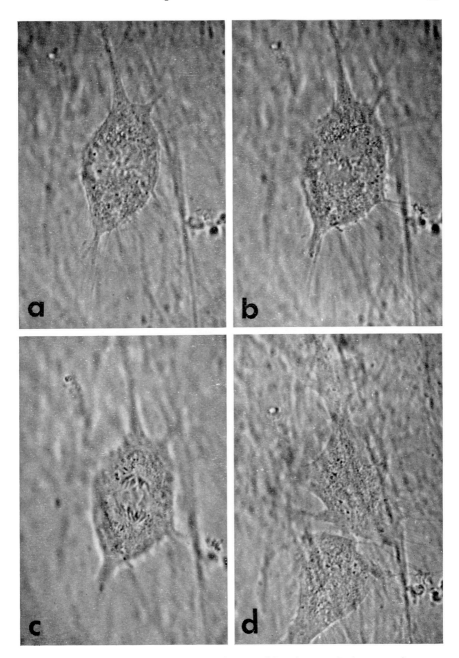

FIG. 2. Stages from a mitotic division in a deuterated (25%) sympathetic neuron from 18-day mouse foetus, 43 days *in vitro*. (a) Prophase, 0 hr. (b) Metaphase 1 hr. 12 min. (c) Anaphase 1 hr. 30 min. (d) Interphase 2 hr. 40 min.

the spreading of the neuronal area cannot be attributed simply to migration; there is an expansion of the central units which forces the whole mass outward. Large mature neurons have been seen from time to time in juxta-

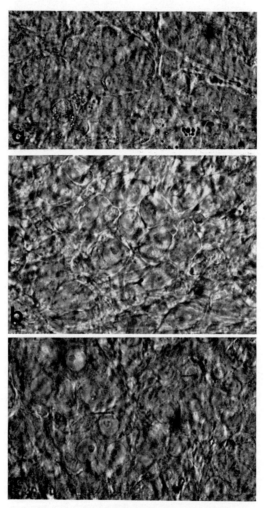

FIG. 3. Sympathetic ganglion from 14-day chick embryo, living. ×450. Eight days *in vitro*. (a) BSS control. (b) NGF, 2 units/ml. (c) 25% D_2O.

position with several neuroblasts or a neuroepithelial cluster which were not observed there the day before. This suggested to us the possibility that nuclear subdivision had occurred within the cytoplasmic mass of a large sister neuron. We have observed one mitosis which resulted in the reconstitution of three nuclei within a large cell; the chromosomal distribution did not seem to be equal, and it was prolonged beyond the usual time. An apparently normal neuronal division has been photographed from metaphase to completion (Fig. 2). Whatever may be the mechanism of their origin, the incidence of clusters of small, undifferentiated neurons increases throughout the prolonged culture period in D_2O, their compon-

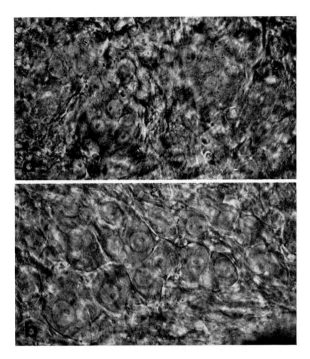

FIG. 4. Same as Fig. 3, 12 days *in vitro*. (a) H_2O control. (b) 5% D_2O.
All photographs in Figs. 4–7 are at the same magnification: × 525.

ent units gradually developing in size and maturity as the cell population spreads peripherally in all directions (Fig. 1, Table I). A good deal of circumstantial evidence suggests that the replicatory phase which is normally confined to early stages of neurogenesis is being prolonged or is recurring cyclically in individual neurons after periods of abnormal growth (Figs. 3–7).

Withdrawal of D_2O from a series of cultures which had been receiving it for 42 days did not result in retrogression or further expansion: the cell populations remained static for 60 subsequent days, after which time the experiment was terminated. *Dissociated* ganglia maintained for 40 days achieved a status more nearly normal in 12·5 per cent D_2O than in other concentrations or in control media.

Living deuterated ganglia appear more opaque and at the same time richer in refringent particles than untreated ganglia. Neuronal nuclei

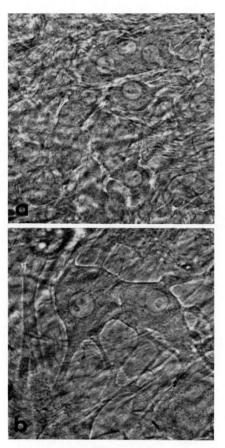

FIG. 5. Same as Fig. 3, 33 days *in vitro*.
(a) H_2O control. (b) 5% D_2O.

especially are turgid and greyish in cast as compared to the clearer, greenish control nuclei. Nucleoli are larger and more prominent in deuterated neurons, where they sometimes may be found as multiple bodies. Arc-ing, rod-like structures may also be seen with greater clarity and frequency in the nuclei of deuterated chick neurons versus controls (Fig. 8). Electron microscope studies of similar D_2O-treated and control chick sympathetic ganglion cultures (conducted by E. B. Masurovsky) confirm and extend many of these light microscopic observations, and reveal other noteworthy fine-structural differences between deuterated and regularly maintained cultures.

Probably the most striking and consistent divergence in ultrastructure that is observed between treated and untreated cultures is the unusual

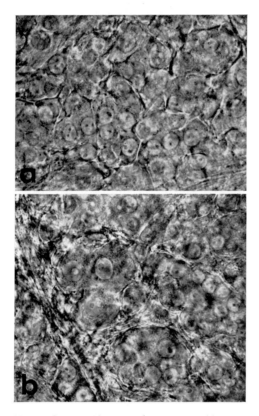

FIG. 6. Same as Fig. 3, 54 days *in vitro*. (a) H_2O control. (b) 5% D_2O.

abundance of fibrous and granular elements in the *nuclei* of the deuterated neurons and supporting cells. Significantly, the distribution and type of such components is variable, possibly reflecting different phases of nuclear synthetic or regulatory activity. Especial interest attaches to prominent bundles of fibrils of about 60 Å, seen at the light microscopic level as linear structures (Fig. 8) which sometimes extend entirely across the

6*

nucleus of deuterated neurons (Fig. 9A). Only relatively sparse fibrillar formations of a similar nature are seen in control neutrons (Fig. 9B). The detailed fine structure and the possible significance of these fibrillar formations are discussed elsewhere (Masurovsky, Benitez and Murray, 1967).

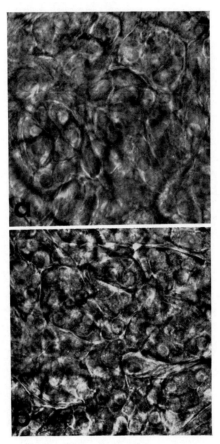

FIG. 7. Same as Fig. 3, 76 days *in vitro*. (a) 5% D_2O. (b) 25% D_2O.

FIGS. 3–7 show particularly well the characteristic groupings of several neurons of similar developmental stage, within capsules.

Nucleoli within these deuterated nuclei appear relatively compact and spheroidal, rather than open-textured or skein-like as is usual in control neurons. The functional significance of this nucleolar configuration is not at present known.

Within the *cytoplasm* of D$_2$O-treated chick ganglion neurons (especially at a concentration of 25 per cent D$_2$O) sheaves of intertwined, branched and beaded fibrils (about 90 to 120 Å), not seen in control neurons, are observed in juxtaposition with elements of numerous, well-developed

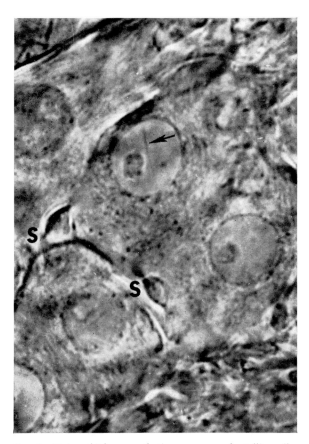

FIG. 8. Living chick sympathetic neurons, and satellite cells, cultured 54 days in medium containing 25% D$_2$O. Note the rod-like structure (arrow) crossing one greyish-appearing nucleus in the vicinity of a prominent nucleolus. Numerous refringent particles pervade the burgeoning cytoplasm of these cells. S: satellite cells. × 1,600.

Golgi complexes and granular endoplasmic reticulum formations (Fig. 10). Microtubules traversing the perikaryon and the branched dendritic-neuritic processes which sprout from these cells appear to be more numerous in deuterated cultures. They are also abundant in satellite and Schwann cells. Unusually dense mitochondria are sometimes observed in deuterated neurons, as are sporadic vacuole and lipid formations. Typical synaptic configurations, containing aggregations of light- and dense-cored vesicles, are found in both D_2O-treated and control cultures, especially those $\geqslant 60$ days *in vitro*.

Salivary glands. Parotid gland rudiments flourish and continue their organoid development after explantation in 5 and 25 per cent D_2O. At a 25 per cent concentration the explanted submandibular gland produces

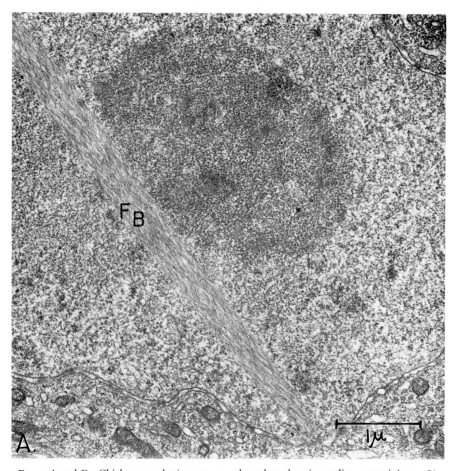

FIG. 9 A and B. Chick sympathetic neurons cultured 54 days in medium containing 25% D_2O (in A), or an equivalent amount of distilled H_2O (in B). The nucleoplasm of the deuterated neuron is marked by an unusual abundance of granular and fibrillar elements. Rod-like intranuclear fibrillar bundle indicated at F_B. Note the relatively compact nucleolus in the deuterated versus the distilled H_2O control neuron. Cultures fixed in Veronal acetate-buffered 2% OsO_4 (pH 7·4). ×21,500.

no outgrowth except possibly a few neurites (from ganglion cells normally present in it), and its cells die *in situ* within two weeks. Five per cent D_2O is better tolerated, though the surviving population is sparse and produces no outgrowth.

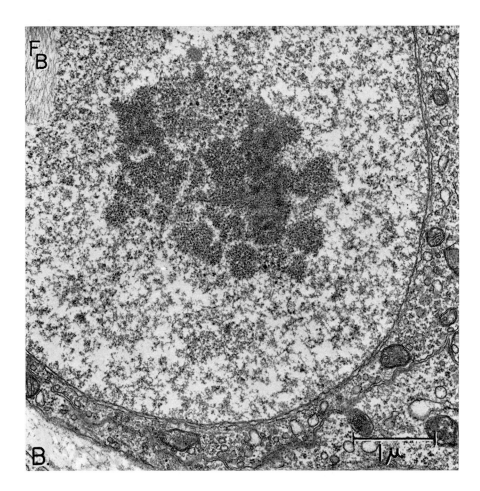

Dorsal root (sensory) ganglia

Cultures of dorsal root ganglia have the advantage of a microscopic visibility which is superior to that of other types of explants in the living state; the ganglion can be set out as a whole, the supporting tissue associated with it is not over-abundant, and the whole complex spreads and flattens. Preliminary observations indicate that growth and multiplication of

Schwann cells, capsule cells and the endoperineurial component constitute the most conspicuous response of this ganglion to D_2O. Endoperineurial cells become especially large and lush. Capsule cells, which should embrace the neuron perikaryon tightly, enlarge and multiply; in so doing they seem

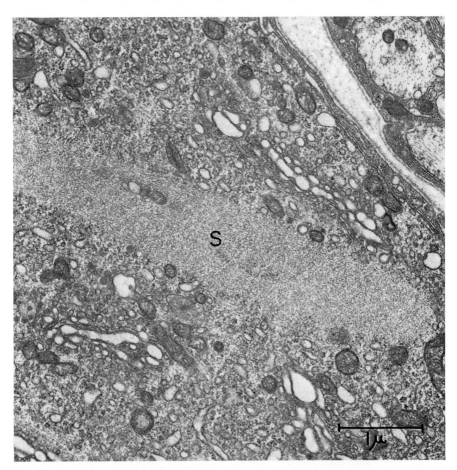

FIG. 10. Cytoplasm of a deuterated (25% D_2O) chick sympathetic neuron containing a voluminous sheaf of intertwined, branched, and beaded $\sim$ 90–120 Å fibrils (at S) not seen in control neurons. Clusters of ribosomes, Golgi elements, and mitochondria appear nearby. Culture fixed in Veronal acetate-buffered 2% OsO_4 (pH 7·4). ×21,500.

to lose their affinity for the neuron, and when it is abandoned by them it deteriorates. Clusters of fleshy capsule cells develop in interstices and appear to push the neurons apart. Neurons themselves initially increase in size in 5 to 25 per cent D_2O; later they become turgid and often lobate; in a few cases they appear to have subdivided, but such a process has not actually been observed. By seven weeks in D_2O there are many (30 to 50 per cent) dead neurons; survivors may appear in good health. The normal mortality is about 15 per cent and this occurs in the first two to three weeks. Myelin sheaths develop in rodent ganglia, and even show rippling overgrowth during the early stages of D_2O administration (three to four weeks), but they become sparser later; this is the reverse of normal culture development. The Y-formed bifurcating neurites are extraordinarily tortuous and branching; large supernumerary neurites are also seen. Ten per cent D_2O is generally well tolerated and produces fewer abnormalities than 25 per cent.

Hypothalamus

In the hypothalamus both neurons and glia respond to D_2O by growth and multiplication. A neuron (16 days *in vitro*) with two satellite cells attached was photographed from early anaphase to the separation of daughter cells in beginning interphase (Figs. 11, 12). The satellites did not participate in any observable way in this division, but remained relatively immobile, one embracing each of two neuritic processes which were never withdrawn. In this area of the brain, number and size of neurons, as well as profusion of neurites, were very much greater in deuterated cultures than in controls (Figs. 13, 14). Since the hypothalamic region has no clearly demarcated anatomical boundary, neurons from adjacent brain areas are sometimes included during the free-hand dissection of small embryos under simple binocular magnification. Myelin sheath formation is then observed in both groups, appearing more profusely in the experimental group. The most favourable D_2O concentration for this brain area is 12·5 per cent.

Cerebellum

Preliminary short-term observations by Dr. C. D. Allerand on a small number of newborn mouse cultures indicate that D_2O has a generally stimulating effect on the cerebellum, and that 10 per cent is more favourable than 25 per cent. Neuron somas mature more rapidly and grow larger than in the controls (Figs. 15, 16); neurites are more numerous and myelin sheaths appear earlier, on more fibres of larger calibre. Stained preparations suggest that all cells, including ependyma and glia, are larger and that the neurofibrillary content of neurons is increased. After three weeks, the beating of ependymal cilia is greatly slowed in D_2O; normal beating, which in the controls is too rapid for the eye to follow, is replaced by a jelly-like quivering.

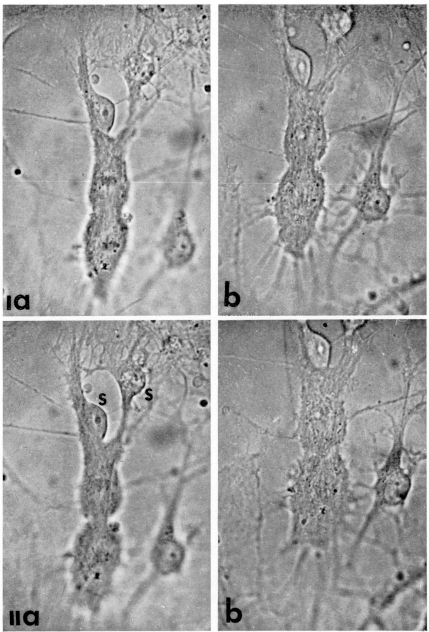

FIGS. 11, 12. Stages from a mitotic division in a deuterated (25 per cent) hypothalamic neuron from 18-day mouse foetus, 16 days *in vitro*. Note satellite cells embracing the two processes of the upper daughter neuron. Another satellite cell has processes which seem to be in contact with the surface of the lower daughter neuron.

I. Anaphase. (a) 0 hours. (b) 7 min.
II. Telophase. (a) 20 min. (b) 30 min.

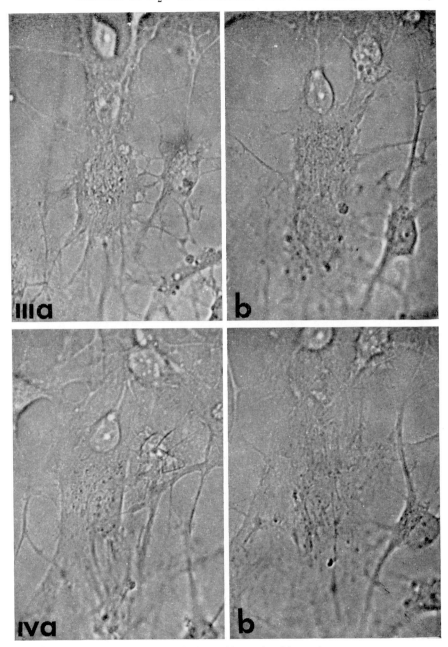

III. Continued telophase. (a) 58 min. (b) 44 min.
IV. Nuclear reconstitution. (a) 3 hr. 20 min. (b) 5 hr.

Because of the depth of this preparation, we have focused on the upper daughter cell in (a), and the lower in (b).

Cerebrum, frontal area

Here, as in the cerebellum, we are dealing with a variety of neuron types and glial relationships, but we are as yet less familiar with their special manifestations in culture. Therefore only crude generalizations can be made at this stage of the investigation.

Within the first three days *in vitro* a felt-like neuritic outgrowth proliferated, especially from the cortical surface. In both 10 per cent and 25 per cent D_2O this outgrowth was heavier and more extensive than in water or in BSS, as was the neuritic branching within the explant. It continued so for the duration of the experiment.

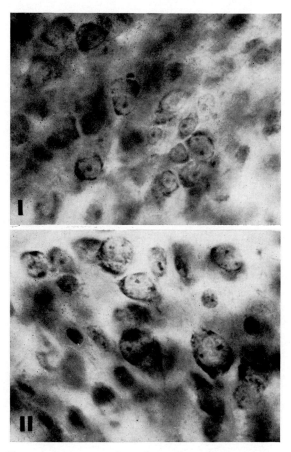

FIG. 13. Twenty-one-day cultures from the same hypothalamic nucleus (on opposite sides) of an 18-day foetal mouse. Jenner-Giemsa stain. ×700. I: H_2O control; II: 12·5 per cent D_2O.

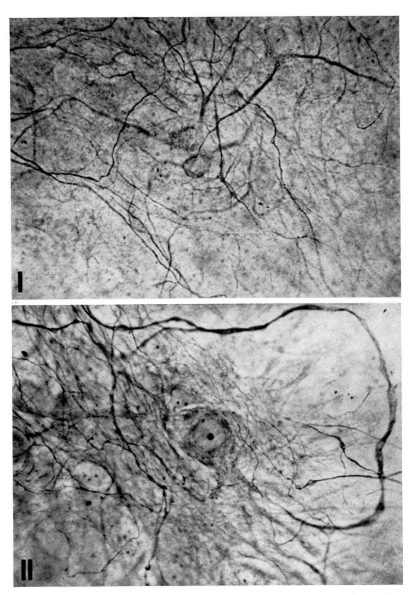

FIG. 14. Fifty-five-day cultures from opposite sides of the hypothalamus of an 18-day foetal mouse. Bodian silver impregnation. ×700. I: H_2O control; II: 25% D_2O.

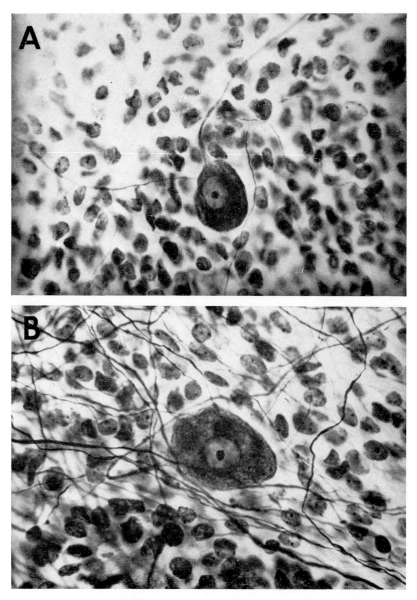

Fig. 15. Newborn mouse cerebellum, 21 days *in vitro*. Holmes, reduced silver nitrate impregnation. ×600. Cells of mesencephalic V nucleus. A: H_2O control; B: 25% D_2O.

After four to five days *in vitro*, neuron somas of the neocortex became distinguishable in all groups. The external granular layer, densely populated with small neurons, appeared brighter in D_2O-treated cultures because of a greater refractivity which characterized their nuclei. Normally large neurons in the deeper layer were somewhat larger in 10 per cent D_2O than in controls; in 25 per cent D_2O some nuclei were eccentric,

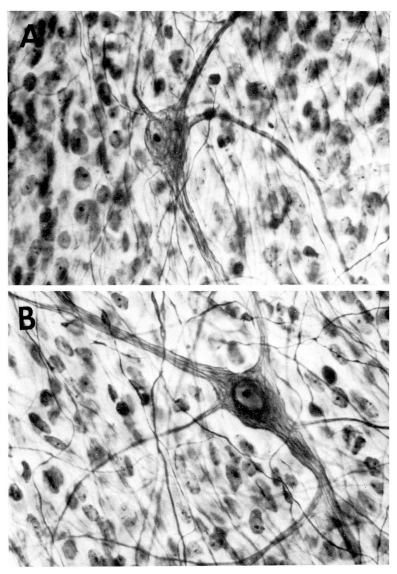

FIG. 16. Same as Fig. 15. Cells of the roof nuclei. A: BSS control; B: 25% D_2O.

and by 40 to 50 days *in vitro* these neurons were rather variable in size. Neuronal mortality remained comparably low throughout in both control and experimental preparations.

From the subcortical region, glial migration was greater and more widespread than from the cortical surface. In D_2O the glial cell population was denser than in controls and exhibited an increased mitotic activity. Central necrosis of the explant, which is a regular occurrence in control media, was practically absent in 25 per cent D_2O, and though present in 10 per cent D_2O it was still of less extent than in the controls.

The first myelinated fibres appear between seven and ten days after explantation, and the peak period of myelination is ten to 24 days. At the beginning, myelination was earliest and heaviest in 10 per cent D_2O. Later, the control groups caught up, so that within 24 days *in vitro* these three groups showed myelination in 90 to 100 per cent of cultures. In 25 per cent D_2O, myelin sheaths first appeared on the tenth day, and the rate and extent of myelination lagged throughout the experiments, although the general health of this group was good. A slow demyelination began in 25 per cent D_2O during the fifth week, as evidenced by herniation and eventual fragmentation of the sheaths.

In cultures fixed at five to six weeks and impregnated by the Holmes reduced silver-nitrate method, deuterated explants show somewhat larger and more fibrillar neurons, with exceedingly long and redundant neurites. Nucleoli, which normally are large but single, are sometimes double or triple. Glial cells have larger and more prominent nuclei than in control media, and unusually fleshy, red-staining perikarya. They do not appear to have deserted the neurons (as in sensory ganglia); indeed it is not unusual to see several satellite nuclei clustered over a neuron soma. It is of course possible that D_2O by its very presence influences the extent of silver impregnation, which is as yet a purely empirical method of delineating various neural structures. Nevertheless, these histological observations are generally consistent with those that we report from cultures in the living state. Ultrastructural studies are in preparation.

DISCUSSION

Numerous and varied biological effects of the deuterium isotope have been described in reports published during the 35 years since its discovery, but relatively few of these studies have included hydrogen replacement in the higher animals, and among these, responses of the nervous system have received still less attention. It has become apparent that, in general, mammals cannot tolerate concentrations of D_2O as high as those to which plants and micro-organisms can be habituated, and that in various cytological parameters the manifestations of replacement may differ also (Flaumenhaft *et al.*, 1965). As yet there are few ultrastructural data available on deuterated animal cells. In our observations of differentiating mammalian and avian tissues exposed for long periods to tolerable D_2O concentrations, some new types of data are presented on the histology and fine structure of the developing and maturing nervous system as affected by the deuterium isotope. It is hoped that these data will afford an entering wedge for more precise investigations of hydrogen transport and biosynthetic pathways which appear to be unique to nervous tissues.

Previous studies on cultured cells have been concerned mainly with established cell lines (HeLa, L cells, monkey kidney, KB cells, etc.); primary rabbit fibroblasts, mouse mononuclear leucocytes and chick embryo cells also have been grown briefly in media containing D_2O (cf. discussion in Flaumenhaft *et al.*, 1965). In these experiments, as in our own, 30 per cent D_2O seemed to be the upper limit tolerated for growth; it was also found (Fischer, 1936) that suffering embryonic cells, if returned to normal medium after 72 hours in higher concentrations of heavy water, recovered completely their capacity to multiply.

Established cell lines are *ipso facto* aneuploid, even though they may not be neoplastic in origin (Hayflick, 1965), and they tend to show abnormal mitotic behaviour. In such monolayer cultures the mitotic rate is generally depressed by D_2O, and a superabundance of giant multinucleate cells may be accumulated, somewhat as in radiation effects. Mitoses do occur in HeLa cells immersed in 20 per cent D_2O, though their incidence is relatively low (Manson *et al.*, 1960). Recovery time for HeLa cells is proportional to the excess D_2O concentration (Siegel *et al.*, 1960).

Lavillaureix (1961) reported that D_2O appears to block mitosis (in KB cells) at prophase and metaphase, and possibly to accelerate anaphase. If so this would fit rather well with the observations of Inoué and Sato (1967) on deuterium effects upon dividing sea-urchin eggs. They note that this isotope increases the volume, length and birefringence of the spindle, encountering different sensitivities at different mitotic stages: a rounded sensitivity peak occurs in prophase and a higher, sharp peak at the end of metaphase. Their electron micrographs indicate that D_2O induces a very rapid and voluminous emergence of microtubules and microfibrils in the spindle area. These authors suggest that the deuterium isotope acts to shift the equilibrium of an already existing pool of spindle proteins towards formation of polymer which is visualized as spindle microtubules or filaments.

It may therefore be significant that we observe an abundance of microtubules in our deuterated nerve tissue. Whether these, appearing as they do most prominently in neurites, are identical with spindle microtubules is open to question. Microtubules which are present continuously as skeletal elements may differ in the details of their origin. In this connexion it might be remarked that in deuterated chick cardiac muscle cells explanted close to sympathetic ganglia and apparently innervated by them, few if any microtubules are seen, though neurites coursing between the groups of muscle cells are richly endowed (Masurovsky and Benitez, 1967; also unpublished observations). These deuterated muscle cells contract strongly and rhythmically; mitoses, with a full complement of spindle microtubules, are seen among them.

Returning to problems of mitosis, one should recall that, in the higher vertebrates, neurons once developed beyond the neuroblast stage are not known to divide, with the possible exception of autonomic cells (see Murray, 1965). Inferentially, mature neurons cease producing DNA, although they are characteristically prolific in RNA production and utilization. Yet in our deuterated cultures of sympathetic ganglia and hypothalamus we see positive evidence of division in large, multipolar cells with which satellite cells have actually established the adult relationship. It may therefore be of particular significance that the nuclei of deuterated neurons are very densely packed with structured particles, many of which conform to the concept of ribosomes. This condition extends also to the specific supporting cells which, though they generally retain their capacity for division throughout life, normally assume in maturity a stable relationship with neurons, and only multiply in response to trauma or other insults. Schwann and satellite cells continue to divide in deuterated cultures at a higher rate than in controls; this is especially evident in the dorsal root ganglion, where the neuron suffers as a result.

However, in this spectrum of nervous tissues we seem to be observing not only proliferation of normally non-dividing cells, but at the same time extensive cell growth with varying degrees of cytodifferentiation. Such a paradoxical combination of developmental processes tempts the observer to speculate anew on the possible effects of deuterium upon macromolecules, especially nucleoproteins, in a system which has not only its own peculiar physical chemistry, but a maze of biosynthetic pathways characterizing its various regions. Is it possible that the deuterium isotope in this tissue might prolong DNA polymerase activity, as it favours polymerization of spindle proteins in cleaving eggs? Or might it inactivate histones, in either case inhibiting nucleic acid repressors? Might it be acting mechanically at the many available membrane surfaces, to change the sterical configurations of high-molecular compounds? It is hoped that ways of investigating these and other possibilities will be found.

SUMMARY

Nervous tissues isolated from the whole organism by explantation respond to D_2O administered in concentrations of up to 25 per cent by accelerated growth and maturation, and also in some situations by multiplication of differentiated neurons.

Histologically different regions of the neuraxis differ in their response to deuterium. Sympathetic ganglia are the most greatly stimulated: deuteration up to 25 per cent accelerates and increases growth of neurons and favours their repeated subdivision as an abnormally large size is attained. The total neuronal population eventually increases twofold or threefold; in turn the progeny—clusters of small, neuroblast-like cells—enlarge and become multipolar. Electron micrographs show an unusual abundance of fibrous and granular elements in the nuclei of both neurons and supporting cells. Fibrillar bundles, not seen in controls, are found in the cytoplasm of deuterated neurons; mitochondria may be abnormally dense, and microtubules unusually numerous.

In sensory ganglia, growth and multiplication of supporting cells is the most conspicuous response to deuteration. Neurons become unusually turgid, and sometimes lobate. They tend to deteriorate as they are abandoned by the activated satellite cells.

The response of hypothalamus to D_2O involves both neurons and glia. A neuron (with satellite cells attached) has been followed through mitosis at 16 days *in vitro*. Growth of cerebellar neurons, extension of neurites, migration of glial cells and myelination are all accelerated by 10 per cent D_2O during early culture stages. Later the controls catch up. In cerebral cortex, 10 per cent D_2O accelerates maturation; some dedifferentiation follows later.

Acknowledgements

The authors wish to thank other staff members of the Laboratory for Cell Physiology for important contributions to this work: Dr. E. B. Masurovsky for the electron microscopy; Dr. C. D. Allerand, Miss Annelies Herrmann and Miss Joliet Bembry for preparing and observing the cultures of cerebellum, cerebrum and dorsal root ganglion respectively; and Mr. Eric Grave for the light-microscope photography.

Financial support was received from Grants No. NB-00858 and SO1 FR-5395 from the National Institutes of Health; 431 from the National Multiple Sclerosis Society; and Research Career Award 5-K6-GM-15, 372 from the NIH to the senior author.

REFERENCES

Bachner, P., McKay, D. G., and Rittenberg, D. (1964). *Proc. natn. Acad. Sci. U.S.A.*, **51**, 464–471.
Barbour, H. G. (1937). *Yale J. Biol. Med.*, **9**, 551–565.
Barbour, H. G., and Herrmann, J. B. (1938). *J. Pharmac. exp. Ther.*, **62**, 158–164.
Bornstein, M. B., and Murray, M. R. (1958). *J. biophys. biochem. Cytol.*, **4**, 499–504.
Crain, S. M., Benitez, H. H., and Vatter, A. E. (1964). *Ann. N. Y. Acad. Sci.*, **118**, 206–231.
Fischer, A. (1936). *Protoplasma*, **26**, 51–55.
Flaumenhaft, E., Bose, S., Crespi, H. L., and Katz, J. J. (1965). *Int. Rev. Cytol.*, **18**, 313–361.
Hayflick, L. (1965). *Expl Cell Res.*, **37**, 614–636.
Inoué, S., and Sato, H. (1967). *J. gen. Physiol.*, **50**, 259–292.
Katz, J. J., Crespi, H. L., Czajka, D. M., and Finkel, A. J. (1962). *Am. J. Physiol.*, **203**, 907–913.
Lavillaureix, J. (1961). *C.r. Séanc. Soc. Biol.*, **252**, 622–623.
Levi-Montalcini, R. (1964). *Ann. N. Y. Acad. Sci.*, **118**, 149–170.
Levi-Montalcini, R. (1966). *Harvey Lect.*, **60**, 217–259.
Levi-Montalcini, R., and Angeletti, P. U. (1968). This volume, pp. 126–142.
Manson, L. A., Carp, R. I., Defendi, V., Rothstein, E. L., Hartzell, R. W., and Kritchevsky, D. (1960). *Ann. N. Y. Acad. Sci.*, **84**, 685–694.
Masurovsky, E. B., and Benitez, H. H. (1967). *Anat. Rec.*, **157**, 285.
Masurovsky, E. B., Benitez, H. H., and Murray, M. R. (1967). *Proc. 25th Annual Meeting Electron Microscope Society of America*, pp. 188–189, ed. Arceneaux, C. J. Baton Rouge, Louisiana: Claitor's Bookstore.
Murray, M. R. (1965). In *Cells and Tissues in Culture*, vol. 2, pp. 373–455, ed. Willmer, E. N. London: Academic Press.
Murray, M. R., and Benitez, H. H. (1967). *Science*, **155**, 1021–1024.
Siegel, B. V., Lund, R. O., Wellings, S. R., and Bostic, W. L. (1960). *Expl Cell Res.*, **19**, 187–190.
Thomson, J. F. (1963). In *Biological Effects of Deuterium*, pp. 85–112. New York: Macmillan.

DISCUSSION

Levi-Montalcini: Do you know whether deuterium affects nerve cells also in the living organism? What about its mechanism of action? Do you believe that nerve cells which undergo mitotic activity under the action of deuterium already show signs of differentiation such as the outgrowth of an axon?

Murray: Under normal conditions differentiated neurons of higher vertebrates do not divide. But under certain circumstances it is possible for sympathetic neurons with outgrown axons to do so—even in the adult (Murray, M. R., and Stout, A. P. [1947]. *Am. J. Anat.*, **80**, 225). I doubt if this is possible in the hypothalamus. I believe the division of sympathetic neurons in the adults is very rare, and I don't know what types of neurons are involved in it. F. de Castro (1923. *Trab. Lab. Invest. biol. Univ. Madr.*, **20**, 113–208) suggests that "reserve cells" which have never fully differentiated come into play in cases of suspected ganglionic regeneration. The effects of D_2O on the whole mouse *in vivo* have been studied by others and neurophysiological abnormalities have been described. But so far as I know no one has examined the nervous system morphologically beyond P. Bachner's personal communication that it appeared grossly normal at the time his mice died of hyper-deuteration. That is why I brought in the physical data of Katz and co-workers (1962, *loc. cit.*) showing that replacement of normal water occurs at a slower rate in the brain than in somatic organs and tissues. Animals succumbing to excess deuteration do not die suddenly; their body tissues are damaged at different rates: the kidney is affected first, and the adrenal cortex (but not the medulla with its neural affinity) early shows morphological lesions, as do the gonads and the salivary gland. We have explanted the submandibular gland in 25 per cent D_2O, like the sympathetic ganglia; a few neurites grew out from the ganglion cells which are normally present in the gland, the glandular cells died *in situ* and eventually the neurites died with them. We then put parotid glands into 25 per cent D_2O and they flourished. Cardiac muscle also will grow very well and contract extraordinarily well in 25 per cent D_2O.

Outgrowing nerve fibres from deuterated ganglionic explants appear sooner and grow to a greater length than those of the controls; but we do not observe the thick, bushy halo that is typical of NGF.

Crain: Have you done any Nissl or silver stains on these cells during mitosis, or shortly afterwards, to see whether they differ from the non-dividing mature neurons in the culture?

Murray: No, not yet.

Szentágothai: Hypertrophy of the neurofibrillar apparatus—neurofilaments and neurotubules—belongs apparently to the standard stock of structural reactions of the neuron that can be observed under various pathological circumstances and even in biological reactions. An extremely strong hypertrophy of neurofilaments (Gray, E. G. [1964]. *Archs Biol., Liège,* **75**, 285–299; Szentágothai, J., Hámori, J., and Tömböl, T. [1966]. *Expl Brain Res.,* **2**, 283–301) occurs in some synaptic terminals in the early stages of degeneration. This corresponds to the hyper-argentophilia of some nerve elements at the beginning of degeneration or even during tissue disturbance. Although fixation artifacts cannot be excluded with certainty, it is our impression that the number and calibre of neurotubules may vary to a considerable degree, depending on the state of activity of neurons. Obvious histological changes of the hypothalamic ventromedial nucleus are induced by bilateral adrenalectomy, suggesting hyperactivity of the neurons. Neurotubules, quite inconspicuous in the nerve cells of this nucleus under normal circumstances, appear to have become strongly hypertrophic five to ten days after adrenalectomy (Szentágothai, unpublished observation). More reliable information in this field would be of interest with respect to the question of the repercussion of specific function on the structure of nerve elements.

Murray: Deuterium in cells that normally divide—specifically in marine invertebrate embryos—brings about a great increase in the volume, length and refringence of the spindle apparatus. This increase includes both microfibrils and microtubules (Inoué and Sato, 1967, *loc. cit.*). We don't know whether the spindle microtubules are biochemically the same as those that normally exist in nerve fibres; but D_2O has a similar action on microtubules in these two kinds of cells. However, in the cardiac muscle cells which we grow in the same deuterated cultures as the sympathetic neurons we see hardly any microtubules in resting cells: they exhibit only microfibrils (which are thought to represent actomyosin). Yet when these cells divide they have a full array of spindle fibres including microtubules. Inoué's idea is that there is always present in his cells a pool of spindle proteins that can be polymerized to form the spindle apparatus which is visualized as an array of microtubules and microfibrils. A little colchicine will reverse this normal mitotic polymerization. The actions of both colchicine and deuterium are reversible and very rapid in such a rapidly dividing system as the marine embryo. Perhaps in our neurons we are witnessing the polymerization by D_2O of spindle proteins which normally exist there in small quantity. But since the deuterated nuclei are much more densely packed with particles (both granular and fibrillar) than the control nuclei, is it possible that the D_2O has in some way interfered with the action of repressors and released the neuron from its normally operating polymerase inhibitors? As regards a mechanism of D_2O action, this is the chief suggestion I have at present; it is as yet unproved.

Muntz: One effect of D_2O on protozoan *Pelomyxa carolinensis* is a reduced membrane potential. In 40 per cent D_2O Pringsheim Ringer solution, the membrane potential of about 70 mv was reduced by about a third. These were only short-term experiments, and reduction happened very quickly after the animal was put in D_2O.

Murray: An interesting observation relating to the protozoa is that the normal phototropic reaction of *Euglena* ceases in D_2O. The only way of detecting phototropism in *Euglena* is to observe its locomotion, which depends on flagellar movement. We find that when mouse cerebellum has been deuterated in culture for two to three weeks, the ependymal cilia which normally move too fast for the eye to follow, practically stop and show only a jelly-like quivering. The skeletal structure there, as in the flagellum, may be augmented or stiffened like the spindle microtubules, since the ciliary and flagellar rodlets are probably modified microtubules.

Hník: Can nerve fibres regenerate in tissue culture after being crushed? Would it be possible to obtain reinnervation of homologous and alien muscle cells and to see whether neurons in tissue cultures lose their specific properties, or whether they would be rejected? This may be an ideal approach to these problems which so far have been talked about in a hypothetical manner.

Murray: We and others are at present trying to do this.

Kerkut: Dr. E. W. Taylor in Chicago has suggested (1967. *Bull. Neurosci. Res. Program*, April, in press. Brookline, Mass.) that there is some similarity between the proteins in the aster and in the neurofibrils in the flagella—they will all combine with colchicine whereas the actinomycin system will not do so. This might fit in with your observation concerning the biological action of deuterium.

Murray: It has a great variety of actions. We have been using certain specifically organized tissues in isolation. Some of our observations contradict those of others in other systems; for instance, our heart muscle contracts much more rapidly and strongly with deuterium than without it. I believe it has been reported that skeletal muscle behaves in the reverse way, *in vivo*.

Gaze: The method Dr. Hník referred to, of using cultures for examining the specific formation of connexions, has been approached, in passing, by various people, including K. W. Jones and T. R. Elsdale in cultures of explanted amphibian neurula (1963. *J. Embryol. exp. Morph.*, **11**, 135–154). They reported the formation of connexions between neurons and developing muscle cells, with the development of spontaneous twitches only occurring once the connexions had been formed. This seems an admirable set-up for examining the maturation of various specific connexions and I am surprised it hasn't been followed up.

Murray: J. Szepsenwol (1947. *Anat. Rec.*, **98**, 67) did this with the chick embryo. But we need to go over this early work again with improved methods.

Crain: In Jones and Elsdale's work, were the nerve and muscle tissues completely separated at explantation or were they left in their natural tissue framework even though the neurons and muscle cells may not have been connected at that time?

Gaze: They were explanted too early and they were allowed to differentiate *in vitro*.

Crain: With Dr. M. Corner, I have reported on differentiation of neuromuscular tissues in cultures, starting with very early frog neurula tissue explanted together with presumptive mesoderm (Corner, M. A., and Crain, S. M. [1965]. *Experientia*, **21**, 422–424). Neurons and skeletal muscle fibres developed during the first few days *in vitro*, and spontaneous, rhythmical twitching patterns appeared by the fourth to the seventh day. Electrophysiological studies demonstrated that spontaneous (as well as electrically evoked) discharges generated in the cord could trigger muscle contractions. The data suggest that the regular endogenous bursts of muscle twitches observed in these cultures may be triggered by periodic bursts of spontaneous neural activity. These results agree with and extend microscopic observations during ontogenetic development of amphibian neuromuscular systems *in vitro* and also *in situ* (Jones and Elsdale, 1963, *loc. cit.*; Corner, M. A. [1964]. *J. comp. Neurol.*, **123**, 243–256; Corner, M. A. [1964]. *J. Embryol. exp. Morph.*, **12**, 665–671). They also support J. Szepsenwol's conclusions regarding the neural basis of spontaneous muscle twitching in cultured chick embryo cord-myotomes (1946; 1947. *Anat. Rec.*, **95**, 125–146; **98**, 67–85), and they tie in nicely with Professor Hamburger's studies of spontaneous rhythmic activity in chick embryo spinal cord *in situ* (see Hamburger, this volume). Similar studies are also in progress with cultures of 12-day mouse embryo cord-myotomes where neuromuscular transmission has been demonstrated after development *in vitro* of highly organized arrays of myelinated ventral-root axons penetrating into bundles of cross-striated muscle fibres (Crain, S. M. [1966]. *Int. Rev. Neurobiol.*, **9**, 1–43). In all of these studies in culture, the nerve and muscle tissues were explanted together without severing the original connective tissue framework, even though the nerve and muscle cells may not yet have been connected at the time of explantation. Under these conditions, the nerve fibres might be able to form functional connexions with muscle more readily than when an artificial gap exists between the nerve and muscle tissue (as in the experiments with coupled CNS explants which I described earlier). Although functional interneuronal connexions clearly developed after formation of neuritic bridges between various CNS explants, preliminary attempts to present skeletal muscle to spinal cord explants have been less successful. Regular formation of neuromuscular junctions may be interfered with by aberrant growth of connective and meningeal tissues. There may also be factors which limit the ability of CNS neurons and muscle cells to make such contacts under these completely isolated conditions. The technical problems impeding use of this model system are now being studied in a number of laboratories, as Professor Murray pointed out, and it is hoped that practical, reproducible techniques for preparing ordered arrays of interconnected nerve and muscle cells will soon become available.

ENDOCRINE INFLUENCES IN NEURAL DEVELOPMENT†

Jerry J. Kollros

Department of Zoology, University of Iowa

A HORMONAL influence in neural development has been recognized at least since the time it became understood that cretinism is dependent upon a failure of normal development of the thyroid gland. However, the relationships of endocrine glands to development of the nervous system were relatively little emphasized in research until about 25 years ago. Since then an increasing concern has been shown, and many organisms at many developmental stages have been investigated with a host of research techniques.

Because thyroid hormones play so dramatic a role in the many processes directing amphibian metamorphosis, it is not surprising to find that they modify neural development as well. But it may be surprising to see just how numerous and diverse such influences are within the nervous system itself. Among the early records are the papers of Allen (1918, 1924), Cooksey (1922), and Schulze (1924), concerned largely with gross morphological features of the brains of frog and toad tadpoles. Allen and Schulze both stress the larval appearance of the brains of animals without thyroid or pituitary glands, particularly the larger size of the ventricles, the thinner walls of cerebrum and medulla, and the breadth of the optic lobes and the medulla, as compared to metamorphosed control animals. In contrast Cooksey and Schulze reported that thyroid feeding accelerated the transformation of gross brain characteristics from those of a larval type to those of an adult. Distortions dependent upon changes in brain-case size were also noted.

Later studies are generally characterized by a more restricted focus, and greater detail. An example is the analysis of the relationship of onset of the corneal, or wink, reflex to metamorphosis. Several earlier investigators,

† Supported in part by grant AM 02202, from the National Institute of Arthritis and Metabolic Diseases.

in ancillary observations, had indicated that the corneal reflex in both urodeles and anurans was absent during all or the major part of the larval period, appearing at or just before metamorphic climax (reviewed in Kollros, 1942). In several species of *Ambystoma* and *Triturus* the reflex in well-fed animals first appears after 70 to 90 per cent of the larval period has been completed; in poorly fed ones it may appear well before 50 per cent of the larval span has passed. Of greatest importance, however, is the onset of the reflex in hypophysectomized larvae, which never metamorphose; in these the onset corresponds in time to that seen in normal larvae of the same growth rate. In urodeles, therefore, metamorphosis is not intimately related to the development of the reflex.

In sharp contrast, however, is the situation in anurans, in which the reflex appears to be very closely related to climax events in metamorphosis. In *Bufo fowleri* and *Rana pipiens* the average time of onset is two to four days before forelimb emergence. In several hylids the onset uniformly follows forelimb emergence by a day or two. Only in a few exceptional specimens of *Rana catesbeiana* did the first indication of the reflex precede forelimb emergence by more than two weeks (Kollros, 1942). Initial attempts to modify the time of onset in relation to forelimb emergence involved stimulation of metamorphosis by immersing tadpoles in strong solutions of thyroxine. In all cases forelimb emergence was advanced, as was onset of the corneal reflex, but the degree of advancement was generally greater for forelimb emergence. In *Rana pipiens*, for example, the first appearance of the corneal reflex, instead of preceding forelimb emergence by an average of four days, preceded it by an average of just two days, or one; in the cases of greatest stimulation, it even *followed* forelimb emergence by one to five days.

These results, though confirming the connexion between metamorphosis and onset of the corneal reflex, permitted no resolution of the question as to causal relationships. Was the influence of thyroid hormone directly upon the reflex centres in the brain, or was the effect indirect? This question was resolved by the implantation of small pieces of agar, soaked in thyroxine, into the fourth ventricle. In tadpoles containing such implants general metamorphic changes were initiated within a few days after the operation, but of special import was the degree of change of the reflex in relation to forelimb emergence, since the reflex now preceded opercular perforation by up to 18 days, and in one group by an average of 13 days rather than by four. Apparently the strong local concentration of the hormone in the fourth ventricle initiated a very rapid maturation of the reflex centre. This interpretation was strengthened by the observation that in several instances the reflex developed earlier on one side than the other, in each instance in correlation with the lateral placement of the implant. Control thyroxine-agar implants placed in the coelom gave results comparable to those of the immersion series, namely, corneal reflex onset was not advanced as much as forelimb emergence (Kollros, 1942, 1943).

Subsequent studies of maturation of the corneal reflex have shown that by careful manipulation of temperature and thyroxine concentration, hypophysectomized tadpoles immersed continuously in weak thyroxine solutions can be brought to just that stage of transformation at which the corneal reflex can be elicited, but which represents a barely subthreshold condition for forelimb emergence (Kollros, 1958). Recently this same condition has been achieved in both normal and hypophysectomized tadpoles of *Rana pipiens* subjected to immersion or immersions for limited periods in relatively concentrated solutions of thyroxine (Kollros, 1966). Since the time of that report, separation of over 30 days between onset of corneal reflex and forelimb emergence, or death, has been achieved. It may be possible to achieve corneal reflex onset without at the same time so modifying mouth parts and behaviour as to preclude continued feeding. Hypophysectomized animals might then survive indefinitely as tadpoles while displaying the reflex.

Of some interest may be the role of metamorphosis in changing the character of the skin of the head. In newts this change results in a gradual restriction of the reflexogenous area (Weiss, 1942; Kollros, 1943). This observation has been confirmed by Székely (1959), who further reports that if a limb blastema substitutes for an ectopic cornea just anterior to the gills, a corneal reflex can be obtained temporarily from stimulation of the blastemal skin but not from that of adjacent non-regenerating limb skin, nor from the fully regenerated limb. His interpretation of these results is based on the character of the early-formed sensory endings, and on the organization of the centres to interpret the sensory code received from such nerve endings of a "low grade of differentiation".

A system showing both direct and indirect influence of thyroid hormone is that of the lateral motor column (LMC), i.e. the ventral horn, which provided the motor innervation to limb and girdle musculature. This group of cells is first distinguished from the adjacent grey matter as a result of lateral migration, at about the time 25 per cent of the larval period has been completed. The cells continue their slow growth and maturation, and soon a reduction in their number begins (Beaudoin, 1955). Definitive numbers are apparently established at the time of metamorphic climax, whereas growth of the cells continues well into the post-metamorphic period. It was shown by Beaudoin (1956), in *Rana pipiens*, that maturational changes in the LMC could be achieved either by placing a thyroxine-containing pellet of cholesterol adjacent to the spinal cord (resulting in a disharmony in the two sides of the cord, the more developmentally advanced being the side containing the pellet), or in the hindlimb (resulting in a greater growth and differentiation of the limb on that side, and a concomitantly greater maturation of the ipsilateral cord, so far as the LMC was concerned). It was shown subsequently that the LMC forms in hypophysectomized tadpoles. Thus the early segregation of these specialized cells depends upon neither pituitary nor thyroid hormones. The limbs of such animals, of course, remain small, differentiating only partially, usually to the condition of early digitation (stage VII). The lateral motor column in these hypophysectomized animals remains in the primitive condition appropriate for the limb condition; cells remain small, with relatively little cytoplasm, and the loss of cells so typical of later stages in the normal animal is not seen, no matter how long the animals are kept. As soon as such animals are treated with thyroid hormones the usual maturational changes begin. Cytoplasm increases in amount, and there is a prompt loss of cell number. However, a fully normal pattern so far as cell size is concerned has not yet been achieved. With low hormone concentrations metamorphic progress is slow, but limb development is normal, and large well-formed limbs are seen; nonetheless, LMC cell size increases much less than is normal for the given developmental stages.

When hormone levels are increased, either to moderate or to significantly excessive concentrations, metamorphosis is moderately or greatly accelerated. Limb growth may be abnormal, with differentiation appearing long before the usual amount of growth. Under such circumstances, also, LMC cell size remains well below normal. Correspondence between stage and cell number is either good or excellent, but not that between cell size and either stage or limb size, and obviously not between cell size and cell number (Kollros and Race, 1960; Race, 1961). Shortly after this study, another in our laboratories by Reynolds (1963) demonstrated that potential or prospective LMC cells were capable of responding to thyroid hormone in stages before their expected appearance. First, thyroid hormone treatment of normal tadpoles in stages younger than the time of appearance of the LMC resulted in an acceleration of the appearance of that cell column, although not earlier than the development of the limb to the appropriate stage. Further, still earlier stimulation of the tadpole by thyroxine produced growth of cells in the grey matter adjacent to the expected site of appearance of the LMC, and thus presumably of the presumptive LMC cells themselves quite prematurely. Not only did some cells grow, with nuclei significantly larger than those of the adjacent more central nuclei, but some cells degenerated, a condition never seen in control animals of the same age from the same egg clutches. Thus, both the growth and degenerative responses typical of cells of the lateral motor column could be elicited before the actual period of formation of the column, that is, before lateral migration of the group of neuroblasts. Similar results have been reported in the unusual *Eleutherodactylus martinicensis* by Hughes (1966), in which it is shown that thyroidectomy retards cell growth and differentiation, while treatment with thyroxine accelerates it. Absence of thyroid hormone, as in thyroidectomy or hypophysectomy, reduces the argyrophilia of the nerve cell body. Unlike the situation in *Rana pipiens*, however, cell number is essentially unaffected by thyroxine treatment or by thyroidectomy or hypophysectomy.

Table I

TOTAL CELL COUNTS IN THE LATERAL MOTOR COLUMN IN *Rana pipiens* LARVAE*

Temp. (°C)	Stage X	XIII	XVII	XX
22	4,900	3,475	2,525	2,245
18	5,435	3,785	2,985	2,530
14	6,040	4,700	3,885	3,620
10	6,445	5,840	4,690	
6	6,920	6,080		

* Data from Decker (1967)

One final experiment which *may* relate to hormone concentration has just been completed in our laboratories (Decker, 1967). *Rana pipiens* tadpoles cultured at different constant temperatures between 22°C and 10°C (e.g. 22°, 18°, 14°, 10°) showed, as expected, slower growth and a larger body size. Each step downward in temperature, however, also resulted in larger than normal populations of lateral motor column cells, for all of the stages which were observed (X, XIII, XVII, XX). The differences in cell number became very significant, and at the coldest temperatures the final cell numbers were approximately twice those seen at the warmest temperatures, or than reported earlier with culture near 25° (Table I). In contrast, cell size was unaffected in the LMC. The exact role of body size, as against that of changes in thyroid hormone concentration in the cold, cannot be assessed at this time. Other studies indicate that thyroid hormone thresholds for given metamorphic events are distinctly raised (Kollros, 1961) in the cold, and it is also suggested that in the cold there is storage of neurohumoral secretory materials in the hypothalamus, and a much slower than normal transmission of these to the pituitary gland, in turn suggesting reduced output of the thyroid-stimulating hormone and thus of thyroid hormone itself (Voitkevich, 1962, 1963; Etkin, 1964).

One particularly interesting type of cell whose full differentiation depends upon thyroid hormone in anurans is that of the mesencephalic fifth nucleus, concerned with proprioceptive stimuli from jaw musculature. It was reported in 1950 (Kollros *et al.*) that the growth of these cells responded to thyroid hormone, and in 1952 (Kollros and Pepernik) it was demonstrated that once growth of these cells had been stimulated they required the presence of thyroid hormone to maintain their size; withdrawal of the hormone resulted in shrinkage of the cell. Further experiments permitted it to be shown (Kollros and McMurray, 1955, 1956) that these cells, which appear about the time that larval feeding begins, are essentially non-responsive to even very large concentrations of thyroxine at very early larval stages (in contrast to most other metamorphosing systems), and only gradually develop such responsivity over approximately the first one-third of the larval period. This responsivity is expressed not only in an increase in the sizes of the cells and their nuclei, but also in the average number of cells which can be identified in individual animals. On the average more such cells are present in stimulated animals than in non-stimulated ones otherwise identically matched. After withdrawal of the hormone, the number of cells is again reduced, presumably by the loss of differentiating characteristics by some fraction of the prospective mesencephalic V nucleus cell population.

By the use of thyroxine-cholesterol pellets implanted adjacent to the midbrain unilaterally, it could be demonstrated that these special cells responded directly to thyroxine, since the cells nearest the hormone source grew larger than did comparable cells on the opposite side, or than did other ipsilateral cells at a greater distance from the pellet. Since the size differences could be detected by five days, but not with certainty earlier, it could be assumed either that rate of response was slow, or that threshold requirements were relatively great. That the latter, at least, is true, was subsequently demonstrated by the use of hypophysectomized animals brought very slowly (by immersion in weak solutions of thyroxine for from 47 to 502 days) to stages of metamorphic climax. Even though these were large animals, with very long legs and other physical features characteristic of the subterminal stages of metamorphosis, the mesencephalic V nucleus cells in these animals remained essentially unchanged in size from those of control hypophysectomized animals. Since the highest thyroxine concentration used in these studies was 0·6 μg./l., threshold levels for cell growth must be above this, but below the level of 20 μg./l. found in other studies to be quite effective.

Inasmuch as the earlier studies with withdrawal of thyroxine after a period of treatment had demonstrated a shrinkage of stimulated cells, it was inferred that reduction in thyroid hormone level in otherwise normal tadpoles might give the same result. This inference was tested in eight *Rana pipiens* and eight *Xenopus laevis* tadpoles by permitting normal development to within five to ten days of the time that forelimb emergence might be anticipated, and then placing the animals in solutions of thiourea, of about 0·05 per cent for *Rana* and 0·25 per cent for *Xenopus*, for eight to 35 days. At fixation matched control animals of the same size and stage were selected. For *Rana*, metamorphic stasis was achieved in only four of the eight cases, and reduction in cell and nuclear size was only suggestive ($P=0·05$). In all eight *Xenopus*, however, metamorphic stasis was achieved, but only after forelimb emergence had occurred. In only four of 48 comparisons were the results contrary to expectations, and only in the case of cell numbers were the differences insignificant (Fig. 1). Similar studies involving 14 pairs of metamorphosed *Xenopus*, with lengths of 20 to 65 mm., yielded similar though less significant results. The differences were greater for the seven small pairs than for the seven large ones (Kollros, 1957).

An additional neuron which has been involved in thyroxine-dependent changes is the cell of Mauthner (M-cell), the atrophy of which has been described as a normal consequence of metamorphosis by Stefanelli (1951), and by Baffoni and Catte (1950, 1951). Baffoni and Catte, and Weiss and Rossetti (1951), reported that atrophy of this cell could be accelerated by treatment with thyroxine. Such a conclusion was supported by the work of Pesetsky and Kollros (1956), using unilateral implants of thyroxine-containing cholesterol pellets; but only for those animals which were strongly stimulated was a bilateral difference in nuclear size evident, with the M-cell on the side containing the pellet being consistently smaller than the M-cell on the opposite side. Subsequent work by Pesetsky (1962), however, makes an alternative explanation more attractive. He presents evidence that euthyroid tadpoles strongly stimulated by immersion in thyroxine show M-cell growth, and he suggests that the shrinkage of the M-cell may be accounted for otherwise. Specifically, he proposes "that during the latter portion of the larval period the rate of enlargement of the anuran M-cell may be accelerated by thyroxine and that it may become sensitively dependent upon thyroid hormone for maintenance. With withdrawal of the hormone, the M-cell would shrink (as do the mesencephalic V nucleus cells) until it could no longer be distinguished from other adult medullary neurons, or it might die." Review of the previous studies on the reduction of the M-cell suggests that this interpretation might be applicable to each of them. In a subsequent study on thyroidectomized *Rana pipiens* larvae Pesetsky (1966*b*) records M-cell sizes significantly smaller than those of three control groups, namely, metamorphosing

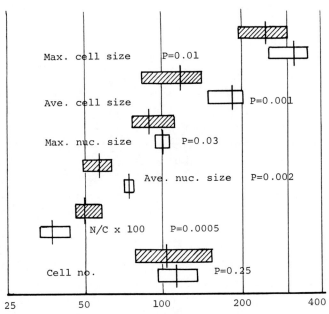

FIG. 1. Characterization of mesencephalic V nucleus cells in eight pairs of *Xenopus laevis* matched for both size and stage. Empty bars represent the ranges of mean values of control animals; hatched bars represent the ranges of mean values of animals maintained in 0·25 per cent thiourea for 14 to 35 days. All thiourea-maintained animals displayed forelimb emergence, followed by metamorphic stasis. The vertical line on each bar represents the mean of means. The scale represents μ^2 for cell size and nuclear size. N/C represents the nucleo-cytoplasmic ratio as obtained in cross-sectional measurements.

animals several days after forelimb emergence (stage XXII), late preclimax tadpoles (stage XVI), and even young larvae of the same developmental stage as his thyroidectomized animals (stage VII) and very much shorter than his thyroidectomized animals (47 mm. as compared with 85 mm.). These observations agree with the notion that the titre of thyroid hormone in the blood is important in the growth and maintenance of the M-cell. Of some interest also is Pesetsky's observation of the much smaller cross-sectional area of the medulla oblongata in the thyroidectomized larvae compared to that of the medulla of smaller but metamorphosing larval control animals. The size discrepancy, which holds for both grey and white matter, confirms the earlier observations of Cooksey (1922) and Allen (1924). Weiss and Rossetti (1951) initially reported that neurons near the M-cell were stimulated to grow as a result of thyroid hormone stimulation, and such growth responses were confirmed by Pesetsky and Kollros (1956) separately for two different size classes of such neurons.

Thyroid hormone has been shown to have an additional influence upon larval and even embryonic brains, namely that of stimulating cell division. This was reported by Weiss and Rossetti (1951) after implantation of hormone sources into the fourth ventricle or adjacent to the choroid plexus in tadpoles. At three days a sixfold increase in mitotic rate in the hindbrain was observed, declining thereafter. Spread of the effect to the spinal cord, particularly the anterior levels, was also seen. It was reported by Baffoni and Elia (1957) that mitotic activity normally increases significantly in both alar and basal plates of the toad metencephalon during metamorphic climax and in the immediate preclimax period. Similar but less striking findings for the alar plate alone of *Triturus* are also recorded (Baffoni, 1957a, b). Stimulation of mitotic activity in the metencephalon of young toad tadpoles by immersion in strong thyroxine solutions was demonstrated by Baffoni (1957c), and interestingly peak activity was shown on the third day of immersion, comparable to the peak stimulation recorded by Weiss and Rossetti (1951) on the third day after thyroxine pellet implantation, and also comparable to the maximum mitotic stimulation reported by Champy (1922) on the ora serrata of *Rana temporaria*, on the third through the fifth days after the start of exposure to desiccated thyroid powder in large amount. In this study Baffoni also reports significant growth of both the alar and basal plates, after both three and five days.

A related study on *Rana pipiens* embryos was concerned with the influence of thyroxine on mitotic activity on both sides of the medulla oblongata after unilateral extirpation very shortly after closure of the neural tube (Ferguson, 1966). One day later, at about embryonic stage 18, immersion in thyroxine was started, and embryos were fixed three to nine days thereafter (four to ten days after operation), at stages 22, 23, 24, 25, and larval stage I. Mitotic density was greater in the thyroxine-treated animals than in the controls at stages 23 to 25, but not at stages 22 and I. Regeneration was virtually complete by stage I, but the total mass of the medulla in experimental animals, considering left and right sides both separately and together, was only insignificantly larger than in the control animals.

Ferguson concludes that "at embryonic stages thyroxine merely accelerates metabolic processes which result in observable increased mitotic activity in cells ready or nearly ready to divide and that upon completion of division a decline in overall mitotic activity follows due to growth and synthetic processes which must occur in daughter cells prior to the next division." The capacity of forebrain and midbrain in anuran larvae to respond by increased mitotic activity to elevated levels of thyroxine has also been shown (Baffoni, 1960).

The use of hormones to produce supranormal growth of the nervous system has been attempted several times. Zamenhof (1941) injected Antuitrin G over periods of 12 to 22 days in *Rana pipiens* tadpoles, without obtaining an increase in body size, but with a significant increase in cell number in the cerebral hemispheres (based on very limited sample sections). Zamenhof later (1942) reported, with more convincing data, increases in cerebral growth in the rat, after injection of Antuitrin G or phyone into the pregnant rat over the period from two to 21 days of pregnancy. Body weights of the litters, at term, were significantly above those of control litters, as were weights of the cerebral hemispheres, volumes of the cerebral cortices, thicknesses of the cortices, total cell numbers, and mean cell densities. By 108 to 124 days of age, the differences between experimental and control groups tended to decrease, with body weights and cerebral cortex weights being insignificantly different, but with total cell numbers and cell density remaining significantly higher for the experimental than for the control groups. Some aspects of this study were duplicated and expanded by Clendinnen and Eayrs (1961), using highly purified bovine growth hormone. They failed to substantiate the postnatal difference in cell density, but indicated a significant increase in the proportion of cortical tissue occupied by perikarya, and in the extent of dendritic development. They also reported some acceleration of behavioural maturation in the experimental animals, and some reduction in errors on the Hebb-Williams closed-field test. The differences between their results and the earlier one of Zamenhof (1942) may perhaps be traced in part to the greater purity of the recent growth hormone preparations, and their lesser contamination with adrenocorticotropin and thyrotropin. Zamenhof, Mosley and Schuller (1966), also using highly purified growth hormone, once again report a significant increase in brain size at birth, and in brain weight: body weight ratios. Further, in the 20-day-old rat they find a significant increase in total DNA content (24 per cent), a much increased cell density (63 per cent), and a still higher difference (71 per cent) in the neuron-glia index. They repeat the earlier conclusion that growth hormone administered to the mother over the major period of pregnancy produces an increased proliferation of cells with neuron-differentiating tendencies. It may be worth pointing out that the foetal brain has not been studied *during* the period of any of these hormone treatments, and thus no estimates of changes in the numbers of dividing cells at any given time can be made.

That other hormones influence behaviour and brain organization has been demonstrated frequently before, but the recent paper by Levine and Mullins (1966) is illuminating in respect to progressive developmental phenomena: it is concerned with gonadal, thyroidal and adrenal hormones, and discusses the importance of levels of these hormones in the few days after birth in determining particular behavioural capacities of the adult. Specifically, critical periods within the first week of life appear to be involved in certain effects of each of these hormone groups. As has been suggested by Eayrs (1964*a*), there may be times early in postnatal life when the nervous system is especially sensitive to thyroid hormone present in excessive concentrations, as well as in deficient ones (Eayrs, 1961). Various aspects of these problems were touched upon by Eayrs (1964*b*) and others at the Ciba Foundation Study Group on Brain-Thyroid Relationships. Relative to the early postnatal critical periods, it is interesting that the primary capillary plexus of the portal system between the median eminence and the pituitary gland is not evident until the rat is five days old (Glydon, 1957). Of equal interest in the studies on amphibians is the information that the median eminence in the hypothalamus-pituitary-thyroid axis develops in the period before metamorphic climax (Etkin, 1965). The dependence of this differentiation upon thyroid hormone has been known, but only recently (Etkin, 1966) has it been demonstrated that the capacity of the median eminence to respond to thyroid hormone arises quite late, and in this sense is much like that of the cells of the mesencephalic V nucleus in anurans (Kollros and McMurray, 1956) rather than like that of most other sensitive tissues (Etkin, 1950; Moser, 1950; Ferguson, 1966); these show responsivity to thyroid hormone at embryonic stages 23 to 25. It will be recalled that Reynolds (1963) demonstrated responsivity of prospective lateral motor column cells at stage 25, so far as cell size increase was concerned; however, onset of responsivity of LMC cells to thyroxine in terms of their degeneration was not established, but it appears to be later than embryonic stage 25.

Specific mechanisms whereby the various hormones exert their effects on the nervous system are largely unknown. Steps towards their understanding include histochemical studies of enzyme localization, such as those of Pesetsky (1965, 1966*a*), enzyme activity studies such as those of

Hamburgh and Flexner (1957), and the studies of Wilt (1959) on changes in visual pigments during metamorphosis. The role of thyroxine or growth hormone in enhancing protein synthesis, or in aiding nitrogen retention, may also be noted. The more general phenomena relating to metamorphosis are reviewed by Weber (1967).

The study of the influences of hormones on the development of the nervous system has revealed a wide repertoire of potential responses inherent within the nervous system. Mitotic activity may be stimulated, growth may be enhanced, or differentiation, or even regression or degeneration. The effects may be evident quickly, or they may be long delayed beyond the time of application of the hormone, and influence maturational events months later. In many instances the role of intrinsic cell or tissue factors must be invoked to explain the onset, enhancement, or disappearance of responsivity, or to explain features of timing, sensitivity and reaction rates in relation to different hormone dosages, times, or modes of application. Not only do hormones influence specific neural differentiation events, but some of these changes in turn influence subsequent hormone production, or release, or feedback adjustments. The absence of a review of these intriguing relationships in amphibians has prompted the emphasis on amphibians here, and the resulting more casual coverage of selected features concerned with mammals, for which some reviews are available.

SUMMARY

The findings of early investigators that thyroid or pituitary hormone deficiencies in tadpoles result in hypoplastic development of various brain parts have been confirmed, extended and particularized. In anurans, in which metamorphosis is hormone-dependent, numerous changes in the nervous system can be shown to be thyroid-hormone-dependent, e.g. the maturation of the centre for the corneal reflex, the maturation of the cells of the lateral motor column (involving degeneration of certain cells, and growth of the survivors), and differentiation of the mesencephalic V nucleus. Studies included premature stimulation of cell maturation by treatment with thyroid hormones, localized treatment by implantation of hormone pellets, delay of maturation through hypophysectomy, thyroidectomy or thiourea treatment, and determination of threshold levels for particular degrees of maturational response, to isolate these aspects of neural development from other metamorphic events. The dependence of the regression of Mauthner's cell upon a fall in thyroid hormone titre was reviewed. The mitosis-stimulating action of thyroxine in brain and spinal cord of embryos and larvae was considered. The capacity of growth hormone to produce excess growth of the forebrain (when administered to the mother during most of pregnancy) in the rat, and to modify behavioural maturation, was briefly reviewed, as were the roles of various hormones acting on the brain in the first week of birth, and influencing behavioural capacities of the adult months later.

REFERENCES

ALLEN, B. M. (1918). *J. exp. Zool.*, **24**, 499–519.
ALLEN, B. M. (1924). *Endocrinology*, **8**, 639–651.
BAFFONI, G. M. (1957a). *Atti Accad. naz. Lincei Rc.*, Ser. VIII, **23**, 90–96.
BAFFONI, G. M. (1957b). *Boll. Zool.*, **24**, 135–144.
BAFFONI, G. M. (1957c). *Atti Accad. naz. Lincei Rc.*, Ser. VIII, **23**, 495–503.
BAFFONI, G. M. (1960). *Atti Accad. naz. Lincei Rc.*, Ser. VIII, **28**, 102–108.
BAFFONI, G. M., and CATTE, G. (1950). *Atti Accad. naz. Lincei Rc.*, Ser. VIII, **9**, 282–287.
BAFFONI, G. M., and CATTE, G. (1951). *Riv. Biol.*, **43**, 373–397.
BAFFONI, G. M., and ELIA, E. (1957). *Atti Accad. naz. Lincei Rc.*, Ser. VIII, **22**, 109–114.
BEAUDOIN, A. R. (1955). *Anat. Rec.*, **121**, 81–95.
BEAUDOIN, A. R. (1956). *Anat. Rec.*, **125**, 247–259.
CHAMPY, C. (1922). *Archs Morph. gén. exp.*, **4**, 1–58.
CLENDINNEN, B. G., and EAYRS, J. T. (1961). *J. Endocr.*, **22**, 183–193.
COOKSEY, W. B. (1922). *Endocrinology*, **6**, 393–401.
DECKER, R. S. (1967). M. S. Thesis, University of Iowa.
EAYRS, J. T. (1961). *J. Endocr.*, **22**, 409–419.
EAYRS, J. T. (1964a). *Anim. Behav.*, **12**, 195–199.
EAYRS, J. T. (1964b). In *Ciba Fdn Study Grp Brain-Thyroid Relationships, with special reference to thyroid disorders*, pp. 60–74. London: Churchill.
ETKIN, W. (1950). *Anat. Rec.*, **108**, 541–542.
ETKIN, W. (1964). In *Physiology of Amphibia*, pp. 427–468, ed. Moore, J. A. New York: Academic Press.
ETKIN, W. (1965). *J. Morph.*, **116**, 371–378.
ETKIN, W. (1966). *Neuroendocrinology*, **1**, 293–302.
FERGUSON, T. (1966). *Gen. comp. Endocr.*, **7**, 74–79.
GLYDON, R. ST. J. (1957). *J. Anat.*, **91**, 237–244.
HAMBURGH, M., and FLEXNER, L. (1957). *J. Neurochem.*, **1**, 279–288.
HUGHES, A. (1966). *J. Embryol. exp. Morph.*, **16**, 401–430.
KOLLROS, J. (1942). *J. exp. Zool.*, **89**, 37–67.
KOLLROS, J. (1943). *J. exp. Zool.*, **92**, 121–142.
KOLLROS, J. J. (1957). *Proc. Soc. exp. Biol. Med.*, **95**, 138–141.
KOLLROS, J. J. (1958). *Science*, **128**, 1505.
KOLLROS, J. J. (1961). *Am. Zoologist*, **1**, 107–114.
KOLLROS, J. J. (1966). *Am. Zoologist*, **6**, 553.
KOLLROS, J. J., and MCMURRAY, V. M. (1955). *J. comp. Neurol.*, **102**, 47–63.
KOLLROS, J. J., and MCMURRAY, V. (1956). *J. exp. Zool.*, **131**, 1–26.
KOLLROS, J. J., and PEPERNIK, V. (1952). *Anat. Rec.*, **113**, 527.
KOLLROS, J. J., PEPERNIK, V., HILL, R., and KALTENBACH, J. C. (1950). *Anat. Rec.*, **108**, 565.
KOLLROS, J. J., and RACE, J., JR (1960). *Anat. Rec.*, **136**, 224.
LEVINE, S., and MULLINS, R. F., JR (1966). *Science*, **152**, 1585–1592.

Moser, H. (1950). *Revue suisse Zool.*, **57**, Suppl. 2, 1–144.
Pesetsky, I. (1962). *Gen. comp. Endocr.*, **2**, 228–235.
Pesetsky, I. (1965). *Gen. comp. Endocr.*, **5**, 411–417.
Pesetsky, I. (1966a). *Anat. Rec.*, **154**, 401.
Pesetsky, I. (1966b). *Z. Zellforsch. mikrosk. Anat.*, **75**, 138–145.
Pesetsky, I., and Kollros, J. J. (1956). *Expl Cell Res.*, **11**, 477–482.
Race, J., Jr. (1961). *Gen. comp. Endocr.*, **1**, 322–331.
Reynolds, W. A. (1963). *J. exp. Zool.*, **153**, 237–249.
Schulze, W. (1924). *Arch. mikrosk. Anat EntwMech.*, **101**, 338–381.
Stefanelli, A. (1951). *Q. Rev. Biol.*, **26**, 17–34.
Székely, G. (1959). *J. Embryol. exp. Morph.*, **7**, 375–379.
Voitkevich, A. A. (1962). *Gen. comp. Endocr.*, Suppl. 1, 133–147.
Voitkevich, A. A. (1963). *Bull. exp. Biol. Med. U.S.S.R.*, **53**, 199–204.
Weiss, P. (1942). *J. comp. Neurol.*, **77**, 131–169.
Weiss, P., and Rossetti, F. (1951). *Proc. natn. Acad. Sci. U.S.A.*, **37**, 540–556.
Weber, R. (1967). In *The Biochemistry of Animal Development*, pp. 227–301, ed. Weber, R. New York: Academic Press.
Wilt, F. H. (1959). *J. Embryol. exp. Morph.*, **7**, 556–563.
Zamenhof, S. (1941). *Growth*, **5**, 123–139.
Zamenhof, S. (1942). *Physiol. Zoöl.*, **15**, 281–292.
Zamenhof, S., Mosley, J., and Schuller, E. (1966). *Science*, **152**, 1396–1397.

DISCUSSION

Eayrs: In opening this discussion I shall try to broaden the field of discourse by commenting on the effects of thyroid hormone on the development of the mammalian nervous system rather than on that of the Amphibia, about which we have heard so much during this symposium. Perhaps the similarities and dissimilarities between the different modes of life and development in mammals and amphibians will become apparent later on in the discussion. Evans's group working at Berkeley, California, during the late 1930s were among the first who drew attention to the great importance of thyroid hormone for those transformations in bodily shape and form, as opposed to growth in size, which we have come to regard under the omnibus heading of bodily maturation. These transformations can be extremely well demonstrated in the rat by chronic dosage of thyroid hormone over the first ten days of the animal's life: it becomes virtually a miniature adult. Such dramatic phenomena tempt one to ask whether any analogy can be drawn between these changes in mammalian forms and the metamorphosis of amphibia. If so, then despite the fact that in mammals there is a change of form without a change of function whereas metamorphosis involves a change of function too, we might expect the nervous system of mammals to be particularly sensitive to the influence of thyroid hormone, as Dr. Kollros has so clearly shown is the case in the Amphibia.

We can make some assessment of the general effect, as Dr. Kollros pointed out, in terms of the behaviour of the individual and of the neurohistology of the nervous system. Innate behaviour, that is to say those responses which appear to be organized irrespective of the influence of environment, is retarded in the absence of thyroid hormone and can be advanced, although certainly not to so great an extent, by giving additional thyroid hormone. Dr. Kollros has previously pointed out, though I now seem to detect some slight reservation in what he has just said, that since these responses do indeed appear in the absence of thyroid hormone, the hormone plays no part in the patterning of the neural mechanisms responsible for their mediation, but rather is concerned in some triggering or facilitating mechanism. It is certainly rather curious that it is in cortically mediated or adaptive behaviour that the influence of thyroid hormone in the mammal becomes so prominent.

We have studied the effect of thyroidectomy upon the forms of an adaptive response at different ages in the rat. The earlier the thyroidectomy, the more severe the effect. After the postnatal age of ten days thyroidectomy produces little or no deficit discernible by the measures used. Thyroid hormone replacement given to neonatally thyroidectomized animals produces virtually the opposite picture. From this we can infer that the central nervous system of the mammal is particularly subject to the influence of thyroid hormone over a period of development confined to the time between birth and about the tenth day of life. The nature of this influence is speculative, but many neurohistological changes certainly occur in the cerebral cortex. These seem to be related not to cell number but rather to the growth of cell processes, because the density of the axonal component of the neuropil and the pattern of the branching of the dendrites in the neuropil are both very significantly reduced and the effect of this is to reduce the probability of interaction between axons and dendrites. When I speak about the probability of interaction I refer to a statistical concept not necessarily related to synapsis, but more recent biochemical studies have supported the view that we may expect synaptic relationships *per se* to be reduced. This could certainly be a factor underlying a diminished responsiveness on the part of the cretinoid animal to its environment and its inferior performance in behavioural tests.

The effect of excess doses of thyroid hormone on the behaviour of the animal later in life does not uphold the earlier promise suggested by an early maturation. A rat given a large dose of thyroid hormone very early in life, followed by no further dosage, looks like a cretinoid animal when adult. The behaviour of such animals both as regards deportment and learning capacity is also considerably impaired by comparison with that of normal littermates. The deleterious effects of giving these doses of thyroid hormone during the formative period are confined to that same period during which deficiency of thyroid hormone is so important. The behaviour of animals given equivalent doses of thyroid hormone on a body weight basis after this critical age shows no significant departure from that of their normal littermates.

It would thus seem that, whereas thyroid hormone in amphibia regulates certain aspects of neural development centred upon the maturation of reflex mechanisms, it exerts in the mammal a significant influence on the cerebral cortex. I should therefore like to ask Dr. Kollros whether he knows of any work in mammals which has been based on his own successful implantation of pellets close to the developing brain in amphibians?

Kollros: I don't know of any such work but I hope there is some.

Eayrs: Have you noted any deleterious effects when thyroid hormone is given in excess to amphibians?

Kollros: When thyroid hormone, whether it be thyroxine or any of the available analogues, is given in excess, the result is surely to engage each metamorphosing system maximally. As a result all the systems proceed from exactly the same starting point instead of being engaged one after the other; they get completely out of phase, and this kind of animal very frequently dies very soon (at perhaps seven or eight days) (Gudernatch, J. F. [1912]. *Arch. EntwMech. Org.,* **35,** 457–483). A single massive injection in large bullfrog tadpoles generally results in a modified but very significant metamorphosis; they appear somewhat more normal than if they were continuously immersed in a very strong hormone solution (Dolphin, J., and Frieden, E. [1955]. *J. biol. Chem.,* **217,** 735–744).

Eayrs: Are you then suggesting there is a differential sensitivity of peripheral tissues to a standard titre of endogenously secreted hormone?

Kollros: Yes, in normal metamorphosis. The hindlimbs, for example, are the most sensitive, and they will respond when the external concentration of DL-thyroxine is 0·002 µg./l., which is a very low concentration. But in normal development one system begins after the other as the increase in thyroid hormone goes up; therefore system A which begins first is well along towards metamorphosis before system F starts. If we suddenly put an animal into a high concentration of hormones, all these systems start from the same zero point.

Szentágothai: Thyroid tissue fragments have been implanted into various regions of the brain in mammals in my laboratory, Professor Eayrs, but with the aim of investigating feedback action of thyroxine on the hypothalamus (Szentágothai, J., Flerkó, B., Mess, B., and Halász, B. [1963]. *Hypothalamic Control of the Anterior Pituitary,* pp. 166–168. Budapest: Akademiai Kiadó). There was no significant effect on the hypothalamus. Thyroid implants in the sella turcica yielded clear effects of depressed thyrotropic activity. Is the deleterious effect you see in mammals caused by feedback action destroying the thyrotropic activity of the hypothalamus or the anterior pituitary, or do you think that it is some direct effect?

Eayrs: I have really no idea what is happening. What you suggest did occur to us and we did certain tests of hypothalamo-pituitary-thyroid function. These showed that there is a reduced resting level of thyroid function. The biological half-life of radio-iodine is increased from six days to seven days while the ability of the thyroid to trap the isotope is reduced, but the hypoplastic response of the thyroid to goitrogen is proportionately the same as that given by the normal gland so that, by that criterion, there is little effect on the hypothalamus. The effect of giving goitrogen on the thyroid/serum ratio in the gland blocked by propylthiouracil is such as to show that the effect is significantly greater in animals receiving early treatment with thyroid hormone than in the normal animal. This suggests that the pituitary is fully capable of releasing thyroid-stimulating hormone and indeed even more vigorously than does that of the normal animal. So I don't think the answer lies there. We may find some sort of indication in a study of cerebrocortical histology. Under the early influence of thyroid hormone perhaps dendrites or other processes may grow out and then regress later. As Dr. Kollros suggested, events may get out of phase so that a distorted structure is left. But this is purely conjectural, of course.

Piatt: Presumably the effect of thyroid on the mesencephalic cells is not specific, Dr. Kollros, and you chose these cells because they are larger and more easily measured. Would the thyroid hormone affect other cells adjacent to the pellet in the mesencephalon?

Kollros: We don't know if it does, but you are right about why we chose these cells. We didn't measure the others and they might have increased too, although there is no evidence that they did.

Piatt: Were you able to see any gradient from more laterally inwards towards the mesencoele, in other words away from the pellet?

Kollros: There was a gradient from one side to the other, but not towards the optic ventricle because the pellet was put in front of the midbrain, rather than moved laterally.

Piatt: Does your lateral motor column correspond to the secondary motor cells, and your mesial column to the primary? The way you described the orientation and position of these two cell groups they would appear to correspond to these two types in the urodele.

Kollros: To the extent that the concept can be used in anurans, I think that would be correct.

Levi-Montalcini: Did you or anybody else study the effect of thyroid hormone on nerve cells *in vitro*?

Kollros: Students in several places, including Mount Holyoke College, are doing *in vitro* work in the frog.

Crain: M. Hamburgh and R. P. Bunge (1964. *Life Sci.*, **3**, 1423) have shown precocious development of myelin in cultures of rat cerebellar tissue after the administration of thyroid hormone. The effect was even more dramatic when they cultured the tissues at a temperature several degrees lower than the normal one.

Levi-Montalcini: Zamenhof described a growth effect of the growth hormone on the cerebral cortex of rat foetuses. We were not able to replicate these results and I wonder whether they were confirmed by others.

Kollros: In Zamenhof's 1941 work (*loc. cit.*) on the tadpole he used only one cross-section of the telencephalon. This is an inadequate statistical sample, so I don't think his data were good. In the rat in 1942 he clearly had an increase in growth in the cerebral hemispheres. In 1966 he also reported growth in the cerebrum. Professor Eayrs has done some work along these lines.

Eayrs: It is very hard for me to comment on Zamenhof's work, because we are in some measure of disagreement. In his 1942 studies (*Physiol. Zoöl.*, **15**, 281–292) he got extremely dramatic effects on cortical thickness, body weight, brain size, and so on. B. G. Clendinnen and I (1961. *J. Endocr.*, **22**, 183–193) were unable to repeat this work but did find changes different from those which he described; there appeared to be no increase in cell density but instead dendritic growth was increased. Our findings stimulated Zamenhof to do his 1966 study (Zamenhof, S., Mosley, J., and Schuller, E. [1966]. *Science*, **152**, 1396–1397), the results of which were certainly less dramatic than those of his earlier work though they were claimed by him to be of sufficient significance to support his 1942 findings. It is perhaps encouraging that where our two studies overlap the trends are similar, though his findings are statistically significant where ours are not. Allowing for a margin of error on both sides the results may not be irreconcilable in the end.

Crain: Although the morphological results in Zamenhof's recent experiments have been somewhat weaker than in his earlier work, he has done some correlative behavioural studies with Dr. Essman (Block, J. B., and Essman, W. B. [1965]. *Nature, Lond.*, **205**, 1136–1137) which appear to support a significant enhancement in learning ability of rats given this growth hormone treatment.

Eayrs: It should be added that Zamenhof studied rats aged 20 days, and ours were adults. This provides a possible explanation for our differences, for after an early hyperplasia of neurons a lot may have disappeared. In other words there may occur a proliferation of functionless neurons which eventually would disappear, but this remains to be investigated.

Stefanelli: In the anurans the Mauthner's cells do not seem to be under the direct control of thyroid. In *Triturus cristatus* I find different developments in the size of these cells with the different seasons. In winter they are smaller and in summer, when the animals are aquatic, they are larger. There is a parallel behaviour of the development of the fin and of the lateral line. Metabolic activity may be much more developed in summer than in winter, and sex hormones may play a role. The Mauthner's cells which persist in adult urodeles might therefore be under a different control than in anurans in which they behave as larval neurons.

Kollros: This is another example of urodeles being less dependent than anurans on thyroid hormone for their maturation, just as is shown by their corneal reflex. In terms of the presumed or potential dependence of Mauthner's neuron upon tail or upon lateral line in the anuran, I would recall that Weiss and Rossetti (1951, *loc. cit.*) said that removal of much of the tail produced essentially no change in the size of the Mauthner's neuron. Pesetsky (1959. Ph. D. thesis, University of Iowa), using some of my material and some of his own in which we had grafted extra heads to the backs of other tadpoles, thus freeing them completely from a trunk and tail, developed Mauthner's neurons which persisted as long as the animal remained a tadpole. I cannot recall whether that particular preparation had lateral lines or not—there were many such specimens, but my impression is that some of these either lacked lateral lines or possessed only fragmentary ones. The influence of these two components, the tail and the lateral lines, on the maintenance of the Mauthner's cell appeared to be perhaps minor in the anuran.

Stefanelli: Another typical larval system in amphibians is the Rohon-Beard cells of the spinal cord. G. M. Baffoni and G. Marini (1967. *Atti Accad. naz. Lincei Rc.*, in press) have recently found that when the tail of *Triturus* larvae regenerates in the new cord these cells differentiate again.

Kollros: Dr. Hughes records the retention of Rohon-Beard cells for a longer period of time in thyroid-deficient or thyroidectomized *Eleutherodactylus*. L. Stephens at Long Beach State College has immersed tadpoles in thyroxine and showed there was accelerated loss of Rohon-Beard cells (1965. *Am. Zoologist*, **5**, 222–223). In anurans some of these cells remain until emergence of the forelimb or very slightly later. But there is a very quick loss of cells between larval stages I and II. There are perhaps 220 cells in the trunk in the larva just as it begins feeding. Five or six days later this number has been reduced to 110 and a month later to 40. After that there is a very slow loss and, as in Dr. Hughes's case, there is an anterior-posterior decline in these particular cells, the last one disappearing only at the very end of metamorphosis (Suter, J. H. B. [1966]. M.S. Thesis, University of Iowa). I just don't know whether there is thyroid dependence here.

Székely: The tadpole has a double layer of cornea: an inner layer, and an outer layer called the skin cornea which is fairly insensitive but which dissolves during metamorphosis. Thyroxine treatment might accelerate the destruction of the outer layer and since the inner layer of cornea is more sensitive this may be why the corneal reflex appears earlier. This alternative explanation might bridge the gap between anurans and urodeles. You mentioned that the corneal reflex is thyroxine-independent in urodeles but not in anurans, but why is that so? In urodeles the corneal reflex appears just the same around metamorphosis in well-fed animals.

Kollros: At metamorphosis in the tadpole the two corneas fuse. Fusion can be accelerated by local implants of thyroxine and cholesterol into the organ (Kaltenbach, J. C. [1953]. *J. exp. Zool.*, **122**, 41–51). Have you seen the outer layer disappearing?

Székely: No, I have not; I only thought that that is how the definitive cornea develops. But anyway the outer layer is insensitive, isn't it?

Kollros: It is not insensitive. If you touch the outer cornea, the tadpole will swim away.

Székely: Have you tried peeling off the outer layer, and then touching the cornea?

Kollros: No.

Hughes: Your very beautiful presentation and far-reaching analyses of hormonal effects in amphibia make one think of the differences there are between one species and another. In *Eleutherodactylus*, which of course is quite untypical, the limbs develop as early as they do in an amniote. They are there for nearly the whole development. In *Eleutherodactylus*, deprivation of thyroid either by thyroidectomy or hypophysectomy, even though these operations were admittedly on the late side, did not affect the normal change in the number of ventral horn cells, which went down at the normal rate, although the histogenesis was altered and the production of ventral root fibres was well below normal. I would have thought that the normal reduction in ventral horn cells was in effect a three-way interaction between thyroid, limb and ventral horn. I thought that the limb being present all the time indicated that the effect on the ventral horn cells was primarily from the limb. In *Rana pipiens* after thyroidectomy the limb doesn't develop. Does the number of neuroblasts in the ventral horn stay high?

Kollros: Yes. The limb remains at stage VII indefinitely and the number of neuroblasts remains high indefinitely. Normally there is almost no reduction before stage X or XI and then a rapid one by stage XIV.

Kerkut: Has anybody done similar experiments in the tadpole to those of G. W. Harris and S. Levine (1965. *J. Physiol., Lond.*, **181**, 379–400) in the rat, where by injecting hormones they can get the nervous system to develop into the male or female pattern?

Székely: There are no male and female tadpoles in this sense!

Kerkut: Witschi was able to get genetically-determined tadpoles all of one sex—whichever one he wanted.

Prestige: Professor Hamburger and Professor Levi-Montalcini in their amputation studies on the chick embryo showed that the limb has two functions affecting the dorsal root ganglion cells, one being to produce more cells and the other to maintain cells. I have done a similar study on *Xenopus*, with similar results. In dorsal root ganglia many cells degenerate in *Xenopus* in normal development, as in the chick embryo. These cells are apparently very sensitive to thyroxine. If one adds thyroxine the number of degenerating cells goes up and if it is taken away by putting the animal in phenylthiourea, the number goes down. Was this due to the thyroxine acting directly on the maintenance of the cells or on the production of cells? In an attempt to answer this we kept the animals in phenylthiourea and after five days no more degenerating cells at all were seen. If the cells can only degenerate when thyroxine is present, cells won't degenerate when the leg is amputated. On the other hand, if the thyroxine is acting in some other way and the cells are independent of thyroxine for maintenance, they would still degenerate. They do still degenerate and thyroxine is therefore probably causing them to degenerate, not by inhibiting the maintenance action but by producing more cells to come into the dorsal root ganglion. There are then too many cells coming in, so that the same amount of maintenance is insufficient. So if one raises the production by thyroxine, the maintenance capacity remaining the same, then the number of degenerating cells is also raised.

Kollros: Have you demonstrated that there is in fact an increased production in the ganglia as a result of thyroid hormone treatment?

Prestige: Only by this argument.

Kollros: Have you checked the actual number of cells dividing during a period of treatment with thyroid hormone?

Prestige: No.

Kollros: I was trying to get information about the specific potential for over-production with thyroid hormone in this case. We already know of its action in stimulating mitosis. Weiss and Rossetti (1951, *loc. cit.*) and Ferguson (1966, *loc. cit.*) in different conditions in very young and midlarval tadpoles of *Rana pipiens* have demonstrated that with thyroid hormone treatment mitotic activity will increase in forebrain, midbrain and hindbrain. W. A. Reynolds (1966. *Gen. comp. Endocr.*, **6**, 453–465) has done the same for the spinal cord. The question now is whether, apart from the actual production of structures which are already there, there is ever an overgrowth? Or is it simply that one cell which is ready to divide is suddenly triggered by the hormone two or three days after the beginning of hormone application? Weiss and Rossetti's data suggest it takes about three days to get a maximum result in the tadpole hindbrain. Champy had this same three-day figure back in 1922 in the ora serrata of the *Rana temporaria* tadpole, and Ferguson had this in the embryonic hindbrain of *Rana pipiens* very recently. The question as to over-production has never been demonstrated, and if you have such a system it would be very fascinating, Dr. Prestige.

PENETRATION OF LABELLED AMINO ACIDS INTO THE PERIPHERAL NERVE FIBRE FROM SURROUNDING BODY FLUIDS[†]

Marcus Singer[‡]

Department of Anatomy, School of Medicine, and Developmental Biology Center, Western Reserve University, Cleveland, Ohio

A unique feature in the differentiation of the neuron is the development of the long cytoplasmic processes which may be measured in metres in large animals. These processes, albeit microscopically thin, constitute in their totality a volume greater than that of the cell body (Bodian and Mellors, 1945). Yet together with the cell body they form a genetic, morphological, and functional cellular unit. And, as in the instance of cytoplasm of all other cells, the axon wastes and dies if separated from the parent body. A major problem in the differentiation of the neuron, in addition to classical ones of axonal growth, connexions with the periphery, and functional elaboration, is the development of an adequate biological system to maintain the distant processes. The present paper is concerned with the nature of these mechanisms as at present conceived and with the possible existence of a supplementary mechanism not acknowledged until recently.

The facts of trophic dependence of the axon upon the cell body and of the morphological isolation of the axon within the tubes of glia and myelin have long pointed to the cell body as the sustaining source of metabolic substances of the axon (review: Singer, 1964). There are other reasons to assert the view that the cell body is the axoplasmic fountain which I shall review in the discussion. Indeed, the idea of a somato-axonal passage is so well established that the cell body is generally conceded by implication or assertion to be the sole source of the metabolic and structural machinery of the axon. Thus the axon is viewed as a closed system with a single distant

[†] Work supported by grants from the American Cancer Society and the National Institutes of Health.
[‡] Guggenheim Fellow, Istituto Anatomia, Università Cattolica, Rome, Italy.

metabolic input. This conception of the relation of axonal cytoplasm to the cell body required from the start the demonstration that substances do indeed move along this path, and secondly that a mechanism for the flow exists. There is now good evidence for both of these requirements (reviewed Singer and Salpeter, 1966c).

The evidence comes from radioautographic and biochemical studies on the delivery into the axon of isotopes incorporated into substances of the cell body (see, e.g., Droz and Leblond, 1963; Droz, 1965, 1967; Taylor and Weiss, 1965; Rahmann, 1965; Ochs, 1966) and of secretory granules of various sorts from the cell body (e.g. Dahlström and Häggendal, 1966); and from lapsed-time cinematography of the living nerve fibre (Weiss and Pillai, 1965)—studies which we have reviewed previously and which were recently assessed in some detail in a symposium on axoplasmic flow (Neurosciences Research Program, 1967).

I shall here review and discuss further the results of our recent work which has led us to conclude that the axon is not a closed system, open only to contributions from the cell body, but that there is a second source which is supplementary. Radioactive isotopes of amino acids, when injected into the body of the amphibian, appear in the Schwann and myelin sheath and in the axon even in cases in which the nerve fibres are separated by transection from their cell bodies. The two important parts of our results as we see them are that substances do penetrate into the depths of the myelin wrappings, and that after this passage they may enter the axon. It appears, therefore, that myelin is not an "inert" insulator, as is often thought, but is biochemically active and active in transport, and that the axon receives an input of metabolic substances from the sheath.

LABELLING OF THE SCHWANN AND MYELIN SHEATH WITH RADIOACTIVE AMINO ACIDS

We recently reported (Singer and Salpeter, 1966a, b, c) that after intro-peritoneal injection of [^{3}H]l-histidine in the urodele amphibian, *Triturus viridescens*, the label appeared within the Schwann and myelin sheath. The localization of the label was determined primarily with electron microscope radioautography. Developed grains were scattered everywhere, over the Schwann cytoplasm and nucleus and the myelin wrappings, superficial and deep. The label did not favour a specific region of the sheath, for example the clefts of Schmidt-Lantermann, nor was it concentrated at the node of Ranvier. The extent of labelling increased within the first day and then appeared to decline but persisted for many days thereafter. The label was not uniformly distributed over the fibres. Some fibres were unlabelled;

in others the sheaths were heavily labelled. The sheath of the entire length of the nerve was labelled but often less so distally than proximally.

In our current work other amino acids are being tested in combination with biochemical analysis of labelled nerves. In addition to electron microscope radioautography we have recently relied heavily upon a light microscope procedure in which the nerve is fixed in osmic acid and then embedded in ester wax for sectioning at 2 to 3 μ (Fig. 1). Moreover, for both electron and light microscopy we have varied the technique of fixation, at first only using osmic acid but in later experiments successively paraformaldehyde and osmium tetroxide (as recommended by Peters and Ashley, 1967). Both techniques gave us similar results. We extended our experiments to other animals, particularly the frog, and observed uptake of the isotope by the myelin and Schwann sheath. More recently we initiated a survey of the uptake of various amino acids and completed a detailed study of penetration of the nerve fibre by lysine. Tritiated lysine moved into the sheaths with even greater facility than did histidine. Within 15 minutes the label was detected over the sheath. By one hour the label was quite heavy and it increased thereafter, rapidly reaching a plateau, and then began to decline after five hours. Figs. 1, 2, and 3 are examples of light and electron microscope radioautographs showing the label in the myelin sheath 24 hours after injection of tritiated lysine. The disposition of the histidine label was approximately the same, and the entire length of the nerve was labelled, some fibres more than others.

Initial results with other amino acids show that tritiated leucine, proline, glycine, and phenylalanine also label the myelin and Schwann sheath, but more slowly and in lesser amount. Finally, in each of the above studies the nerves were cut on one side, yet labelling occurred over the distal pieces as well as over similar regions of the uncut nerve.

THE APPEARANCE OF THE AMINO ACID LABEL IN THE AXON

Simultaneously with its deposition in the Schwann and myelin sheath the labelled substance also appeared in the axon. In our light microscope studies on lysine uptake in the nerves of the newt, *Triturus*, we noted a parallel increase of grains over axon and sheath; but the axon lagged behind the sheath in grain counts per unit area and never reached the concentration seen in the former. Preliminary results with the larger fibres of the frog, *Xenopus*, seem at the moment to show that in some cases the axonal label may surpass in amount that of the myelin even at early times. Abundant axonal label is shown in Figs. 1, 2 and 3. Fig. 3 also shows labelled non-myelinated fibres as well as label in the Schwann cytoplasm of

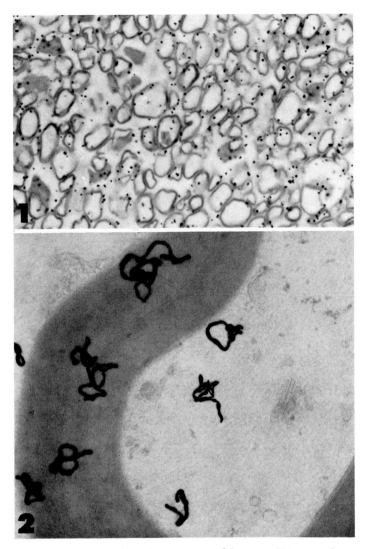

FIG. 1. Labelled nerve from the upper arm of the newt, *Triturus*, 24 hours after transection in the brachial plexus and intraperitoneal injection of 100 μc of [³H]DL-lysine. Fixed in paraformaldehyde followed by 1 per cent osmium, and embedded and sectioned in ester wax. Note label in myelin sheath and axon. × 800.

FIG. 2. Electron microscope radioautograph of nerve of the lower leg from the frog, *Xenopus*, 24 hours after transection of the sciatic nerve and intraperitoneal injection of 2 mc (approximately 50 μc/g.) of [³H]DL-lysine. Fixed in paraformaldehyde, and then osmium; embedded in Epon. Note label over myelin sheath and axon. × 7,800.

myelinated and non-myelinated fibres. Not all axons, as seen in cross-section, were labelled but often only the surrounding sheath; but the frequency of labelled axons increased with time. By five or 24 hours after injection of tritiated lysine or histidine, label was observed in most

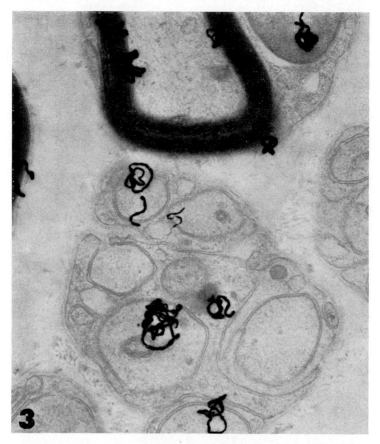

FIG. 3. Another section through the nerve of FIG. 2, showing labelling of the axon and sheath of myelinated and non-myelinated nerve fibres. ×6,000.

of the fibres. Longitudinal sections showed uneven spread along the axon and the sheath; adjacent regions were either bare of grains, or sparsely or heavily labelled. Absence of label in the cross-section of a fibre therefore did not imply non-labelling of the entire axon. Finally, it should be stressed that the label appeared simultaneously along distal and proximal reaches of the axon, although there is some uncertainty at the moment concerning the comparative degree of labelling of these two regions.

Until recently the rate of axoplasmic flow from the cell body distally was generally conceded to be rather slow, of the order of a few millimetres per day (reviewed by Singer and Salpeter, 1966c); this is not fast enough to reach the brachial plexus of the newt within the few minutes or hours of labelling—a distance of about 0·5 cm. distal to the brachial spinal ganglia, or, in the upper arm, a distance of 1 or 2 cm. Samples were taken from the sciatic nerve of *Xenopus* at about 3 cm. from the spinal ganglia. Since much faster rates of flow are now being recognized, it is conceivable that the axonal label may represent a fast component of passage, perhaps followed later by a slower and more massive component. For this reason we routinely transected the nerve of one side distal to the ganglion before injecting the isotope, except in a few instances for a time series study of the early minutes of labelling. Indeed, Figs. 1 to 3 are all taken from nerve distal to the transection. Therefore, whatever flow one may envisage as occurring from the cell body into the axon was interrupted by the cut. There was no evidence that label entered the cut end of the fibre in substantial amount. Consequently, the labelled substance was fed into the axon from the surroundings. We have yet to complete a quantitative comparison of the grain count on the two sides. Our qualitative observations show no obvious differences between the two, attesting to a continuous input whether or not the fibre was separated from the cell body of origin.

Because of the heavy label throughout the layers of the sheath the most reasonable explanation for the appearance of label in the axon is that the amino acid, in whatever form, entered the Schwann and myelin sheath and traversed the space surrounding the axon, whence it penetrated the axolemma to enter the axoplasm. We observed no concentration of label elsewhere, for example at the node of Ranvier, to suggest another source. Our findings give rise to many questions for which at the moment there are no answers. I shall direct the remainder of this work to some of those that are most pressing.

TRANSPORT IN THE MYELIN SHEATH

In our previous publications we suggested a number of possible pathways into the myelin wrappings and thence to the space surrounding the axolemma. I wish to record these thoughts again and to comment further upon them. First of all, substances may penetrate the wrappings to various depths; some are deposited *en route*, to be moved along later, and some are transported without delay to the adneuronal layer of Schwann cytoplasm and thence across the boundaries into the axoplasm. The rapid appearance of the radioactive label in the axoplasm suggests that some component of the movement can be fast.

The pathway into the myelin and the mechanism of the movement may be various. According to our present views about myelin, the wrappings consist of layer upon layer of tightly packed surface membranes. Since surface membranes function in transport, we suggested that such an arrangement of stacked membranes, in addition to whatever other function they may have, may serve to move molecules, large or small, through the thickness of the sheath to various depths. Little is known today about the forces and pathways of movements of charged substances through a solid charged framework. It may be that charged and spatial relations are such within the sheath that certain molecules are accepted rapidly by each layer and passed on to the next. For example, basic amino acids may have more ready access than acidic or neutral ones. They may be bound here or there among the leaflets or otherwise deposited temporarily or for a long time.

Once the transported stuff reaches the internal lamella of Schwann cytoplasm it would now be available to the fluid bathing the axonal surface and the axolemma. The axolemma itself would constitute another selective framework. It may also be that some of the moving stuff is incorporated along the way into the myelin substance, because we observed that the myelin label persisted in the nerve fibre for many days. At any rate such a speculated pathway into the depths of the myelin would be a most direct one.

We also suggested a less direct pathway through myelin, but one which is more easily envisaged on the basis of our present knowledge of the nature of the enclosing sheath. The substances may be transported in a spiral path following the turns of the myelin wrappings, for example within the dense band of the myelin (as seen in osmium preparations) which is believed to represent the compounded inner faces of the Schwann cellular membrane, and therefore the original position of the cytoplasm which was squeezed out during the wrapping process. Thus, instead of traversing the successive faces of the leaflets, the substances would move within the structure of the surface wrapping and travel from the Schwann cytoplasm to the region of the inner mesaxon. The solid framework within this circular path may favour such transport. A lesser possibility which occurred to us is movement within the spiral of the less dense osmicated band representing the space between adjacent extracellular surfaces of the wrappings. However, according to Schmitt (1959) this space is only about 15 Å wide and the surface charges on the faces may very well interfere even with the movement of small ions. A spiral pathway within the internal substance of the wrappings seems more likely when considered in the light of the relations of the wrappings to the Schmidt-Lantermann clefts. Robertson (1962) considers the clefts to be periodic or continuous intrusions of Schwann cytoplasm separating the paired faces of each wrapping all the way to the adaxonal layer of Schwann cytoplasm. The clefts could constitute a spiral cytoplasmic pathway for transport into the depths of the myelin lamellae; and in this case the substances could be carried passively within the stream of cytoplasm. This interpretation of the significance of the Schmidt-Lantermann clefts also led Robertson to suggest that they may serve "for the exchange of compounds between all the membrane surfaces in myelin and the cytoplasmic matrix". Elaborating these thoughts further, one may imagine that the clefts are not fixed in their position within the myelin segment but rather are transient, occurring first here and then there as waves of Schwann cytoplasm advance and recede through the spiralled substance of the myelin wrappings. According to this schema, the Schwann cytoplasm would perfuse periodically and repeatedly the internal substance of the entire myelin lamination all the way to the internal mesaxon and adaxonal layer of Schwann cytoplasm. As the cytoplasm recedes substances would be deposited and captured within the compacting leaflets to be incorporated there or swept along further with the next cytoplasmic wave. Substances that reach the adaxonal cytoplasm would be available for transport to the axon through the intervening space. Should the above-described mechanism of transport in fact obtain, then the myelin wrappings themselves would constitute relatively passive structures of transport. The idea of intruding waves of cytoplasm is all the more attractive in the light of recent lapsed-time cinematography of the nerve fibre, demonstrating pulsations of the Schwann cell body (Ernyei and Young, 1966). Such periodic pulsations may be the mechanism to stir the myelin and to cause the periodic intrusions which constitute the clefts. These pulsations may also be a mechanism for stirring the axoplasm by distorting the enclosed axon.

Thus, the Schmidt-Lantermann clefts themselves provide an obvious cytoplasmic continuum from the Schwann cell body to the surface of the axon. And the thought of periodic intrusions involving different regions of the myelin sheath at different times would not have been invoked in our discussions were it not for the fact that myelin in general, and not clefts alone, is labelled; an explanation would therefore be needed for depositions in these apparently closed regions of the sheath. Yet it is possible that the clefts are fixed in positions from whence substances diffuse into the adjacent relatively closed regions of the leaflet.

The ultrastructural appearance of the myelin sheath is one of stability and the conception commonly presented in textbooks depicts a thoroughly laminated and insulated enclosure of the axon. However, in our experience, particularly with *Xenopus* nerve, many of the fibres do not conform to this

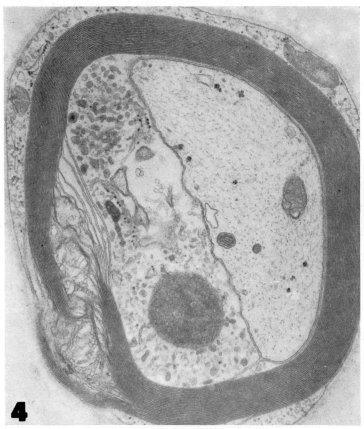

FIG. 4. Electron microscope section of normal sciatic nerve from *Xenopus* to illustrate the frequently observed intimate relation between Schwann cytoplasm and the axon by way of the Schmidt-Lantermann cleft. The adaxonal layer of the Schwann cytoplasm is swollen on the left by a massive cytoplasmic intrusion through the cleft. × 21,600.

common description. Clefts of Schmidt-Lantermann are more frequent than the texts imply and their morphology reveals the existence of a rich cytoplasmic continuum extending through the wrappings from the Schwann cell body to the inner margin of the sheath around the axon. Indeed, at the region of the cleft the adaxonal rim of cytoplasm is often so swollen that it presses the axon to one side. Electron micrographs (such as Fig. 4) which illustrate such massive swelling of the adaxonal Schwann cytoplasm were frequently observed; they show the intimacy of Schwann-axon relations and the fact that at the position of the cleft the axon is distorted

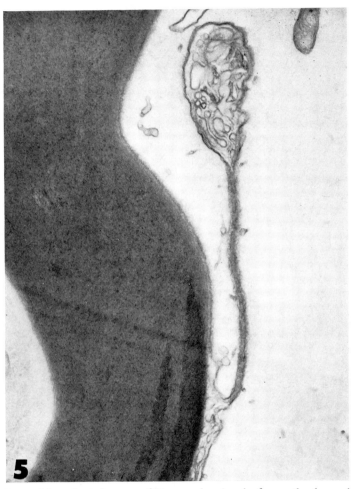

FIG. 5. Normal sciatic nerve of *Xenopus*, showing the frequently observed intimate relation of the axon with the adaxonal layer of Schwann cytoplasm. In this case a balloon-shaped extension of the latter intrudes into the axon. × 21,000.

by the internal swelling. Within the swollen adaxonal cytoplasm there are various organelles including dark bodies and distended tubules as well as surface villus-like folds. The enlargement often presses into the axon itself and one often sees balloon-tipped stems of Schwannian cytoplasm enfolded by the axon (Fig. 5). I emphasize these findings because they show an intimacy between the Schwann cell cytoplasm and the axoplasm which makes transfer of metabolic substances from one to the other quite easy to envisage.

THE NATURE OF THE LABELLED SUBSTANCE OF SHEATH AND AXON

The silver grains of radioautographs bespeak the probable position of radioactive substance. However, in themselves they tell us little about the nature of the radioactive material and how it got there. It is ordinarily accepted that in the procedures of tissue preparation and processing for radioautography, small molecules, including amino acids and peptides, are washed away; but larger soluble molecules are precipitated, and insoluble and structural molecules are also preserved. Indeed, studies by Ochs (1966) and his colleagues have shown that in freeze-substitution procedures which preserve much of the radioactivity in the nerve more label remains than ordinarily appears after common procedures of fixation such as were employed here. In our first biochemical analyses of total radioactivity in formalin-fixed and non-fixed nerves, we observed that approximately 60 per cent of the label is lost immediately after fixation (J. D. Caston and M. Singer, unpublished). Yet there is no real assurance in our procedures and those of others that some of the free labelled amino acid is not bound to high molecular weight substances at the time of fixation. Consequently, I have not employed here the term "incorporated" even though our current biochemical analyses (Caston and Singer, unpublished) reveal measurable incorporation of labelled histidine and lysine into proteins of nerve. Instead, I have used the terms "label" and "labelled substance" and can only speculate about its nature.

Amino acids may be incorporated into the protein of the Schwann protoplasm. Some of the labelled protein may remain there and some may be transported into the myelin layers. Protein incorporation may also occur in the Schmidt-Lantermann cleft and even in the myelin lamellae, from whence it may be carried into the axon. Perhaps small molecules, for example peptides and other derivatives of amino acids, alone or together with larger molecules, may move deeply through the myelin lamellae. Yet it may be that only free amino acids or amino acid associated with a larger molecule penetrates the myelin and axon. It is conceivable that at the time of fixation some of the free amino acid is captured for radioautography by being bound ionically with the highly negatively charged myelin, or by being bound sterically, as Neame (1964) suggests for histidine. Our first biochemical experiments show, indeed, that there is a considerable amount of "free" lysine and histidine and certain of their small molecular derivatives in the nerve 24 hours after injection of the labelled amino acid. As noted above probably more than 60 per cent of the label in the nerve is lost on fixation and subsequent treatment. How much, as seen in the radioautograph, represents entrapped small molecules and how much is incorporated remains to be determined. But whichever it may be, we interpret our results to mean that labelled molecules do enter the sheath and penetrate its depths, and that some are transferred to the axon.

THE RELATION BETWEEN THE SOMATO-AXONAL PASSAGE AND THE INPUT BY WAY OF THE MYELIN AND SCHWANN SHEATH

The thesis that there is a second metabolic input into the axon to supplement the contribution from the cell body does not minimize the importance of the somato-axonal passage. The life of the axon depends on its primary relations to the cell body. However, the thesis does assert that the metabolic input is not confined to the cell body and that the entire length of the neuronal axis among its other functions is available for ingress of metabolites. The two passages into the neuron are probably not equivalent; each may be selective to certain classes of substances. For example, it may be that only small molecules penetrate the axolemma whereas the somato-axonal passage carries a stream of high molecular weight substances from the neuronal perikaryon.

If an interstitial-axonal passage in fact exists, as appears to emerge from our results, this would mean that the axoplasm is not a closed compartment but, like the cytoplasm of cells in general, is open to biochemical messages and contributions along its entire surface. The axon undoubtedly has biochemically unique qualities but so has the cytoplasm of other cells. This thought does not question the biochemical uniqueness of axonal cytoplasm, but it does assert a resemblance between axoplasm and cytoplasm of other cells. There are still other resemblances, for example the presence of organelles found in most cells.

Input of substances directly into the axon without passage from the cell body has another implication, namely that the volume output from the neuronal perikaryon need not be as great as heretofore assumed. Such a passage reduces the burden upon the soma and provides additional assurance that some metabolic needs are met in the distal reaches of the axon. It is conceivable that the relative importance of these two contributions varies per neuron. Some neurons, perhaps according to their size or other structural and functional characteristics, may depend less upon the cell body. This may be why some axons degenerate more slowly than others when separated from their cell bodies. Moreover, the input via the sheaths may account for the lengthy degeneration times observed in transected invertebrate axons. Yet dependence upon the cell body for maintenance of the morphological integrity of the axon is the rule; the axon requires a continuous trophic contribution delivered from the perikaryon with speed because within a short time its absence is reflected in axonal wasting and death.

So far I have discussed only the problem of the metabolic input into the axon and have not considered the source of materials for the cell body. There is growing debate over whether the neuronal soma is fed by the enclosing glial cells or whether it receives its sustenance only directly from the blood stream by way of the extracellular channels (review: Kuffler and Nicholls, 1966).

THE TROPHIC† ACTIVITY OF THE NEURON AND ITS RELATION TO EXTRA-CELLULAR STRUCTURES INCLUDING THE MYELIN SHEATH

Finally, some comment should be made about the trophic dependence of the myelin sheath on the neuron. The present paper has focused upon contributions of the enveloping sheath to the metabolic life of the axon, but the stream is not one way. The facts of Wallerian degeneration reflect a trophic contribution conveyed by the axon to the myelin sheath. Destruction of the axon is followed by degenerative alterations in the sheath; the myelin wrappings break down and disappear and there are profound changes in the cells of Schwann. The changes are reversed and the wrappings are reconstituted once the axon has regenerated along the original pathway. Therefore it appears likely that something is contributed by the axon all along its length to maintain the morphological integrity of the myelin sheath. And it may well be that other substances, including products of metabolism, leave the axon by this route.

The neuron also contributes to the maintenance and the growth of structures upon which it ends. For example, when nerves are cut, muscles waste and taste buds and other organelles degenerate. A dramatic example of the effect of the nerve on peripheral morphology is the inability of amputated body parts in lower vertebrates, for example the salamander, to regrow when the stump is denervated (reviews, Singer, 1952, 1965). The axons pour out an influence—unrelated to conduction as we now know —which maintains and causes to grow the structures upon which they end.

† The term *trophic* as used here defines that quality of the neuron, represented possibly in a neuronal secretion, responsible primarily for the structural maintenance and the growth of the axonal processes but also for the morphological and functional integrity of the tissues upon which the axon terminates and of the myelin sheath which enfolds it.

In previous works I have suggested that the trophic influence is primarily for maintenance of the axonal processes of the neuron; that it probably arises in the cell body and is carried distally in the streaming cytoplasm; and that it is produced in such abundance to spill out of the axon at the endings and therewith to produce its effect upon the periphery (Singer, 1964). To this should be added the possibility that trophic material exudes through the axolemma all along the course of the axon to maintain the morphological and functional integrity of the surrounding sheaths.

THE ORIGIN OF CURRENT VIEWS ON THE METABOLIC INPUT INTO THE AXON

Finally, I wish to summarize some of the more important reasons, as I see them, for the current widely accepted view that the cell body provides the sole metabolic input into the axon. I have commented already on some of the reasons. One is the trophic dependence of the axon upon the cell body, demanding a biochemical continuum or message emanating from the neuronal soma. A second reason, also commented upon, is the constant presence of sheaths which enclose the axon almost completely, and thus provide a structural barrier between interstitial fluids and the axon. The myelin wrappings give the appearance of permanence and stability, and suggest thorough insulation of the axon from events outside the nerve fibre. Another apparent reason is the focus of physiological anatomy, predominantly concerned with biophysical events in impulse conduction and the relation of the sheaths to these events, and much less concerned with the maintenance of the axon and sheath. Also, biochemical studies have indicated that there is little turnover in the lipid of the adult sheath; consequently, myelin is often considered an "inert" barrier (review: O'Brien, 1965) separating the axon from the surroundings, thus affirming impressions derived from cytological and physiological studies.

A historical reason, which indeed may be the ultimate one from which the others stem, is the enunciation of the neuron doctrine, and the debate over its validity, which dominated early neurobiology. The neuron doctrine states that the neuronal cell, including its axonal processes, is a single morphological, genetic, and functional unit and that the axon is not formed by the chains of glial cells which later come to envelop its substance. The long debate over the unit cell hypothesis pre-empted considerations of other possible intimate metabolic relations with the glia. Apparently it was important, in order to establish the validity of this important concept, to ignore and even to deny, at least by implication, a major role to the sheath cells in the economy of the axon. Except for occasional remarks in early and recent literature (see for example Barker, 1899; Schmitt, 1958),

G.N.S.—8

there is little speculation on such a role of the sheath. Since the larger issue of the origin of the axon no longer obscures the problem, it may be time to re-examine metabolic exchanges and interactions between the components of the nerve fibre.

SUMMARY

The present work re-evaluates current views on the relation between the axon and the enveloping sheath. The source of axonal substance is commonly considered to be exclusively the parent cell body; the possibility of metabolic input into the axon from surrounding tissues is largely ignored. There are historical and experimental reasons for limiting ingress to the axon hillock which are discussed here. Recent experimental evidence from our laboratories strongly supports the possibility of a second pathway for substances into the axon. The label of some amino acids was found to penetrate the nerve fibre even when it was separated by transection from the cell body. The label was observed in the Schwann and myelin sheath and in the axon. The significance of these observations is discussed; and the possible nature of the labelled substance and its pathway into the myelin lamellae are assessed.

REFERENCES

BARKER, L. F. (1899). *The Nervous System and its Constituent Neurons*, p. 307. New York: Appleton.
BODIAN, D., and MELLORS, R. C. (1945). *J. exp. Med.*, **81**, 469–488.
DAHLSTRÖM, A., and HÄGGENDAL, J. (1966). *Acta physiol. scand.*, **67**, 278–288.
DROZ, B. (1965). In *The Use of Radioautography in Investigating Protein Synthesis*, vol. 4, pp. 159–175, ed. Leblond, C. P., and Warren, K. B. New York: Academic Press.
DROZ, B. (1967). *J. Microsc.*, **6**, 201–228.
DROZ, B., and LEBLOND, C. P. (1963). *J. comp. Neurol.*, **121**, 325–346.
ERNYEI, S., and YOUNG, M. R. (1966). *J. Physiol., Lond.*, **183**, 469–481.
KUFFLER, S. W., and NICHOLLS, J. G. (1966). *Ergebn. Physiol.*, **57**, 1–90.
NEAME, K. D. (1964). *J. Neurochem.*, **11**, 655–662.
Neurosciences Research Program (1967). *Bull. Neurosci. Res. Program*, April, in press. Brookline, Mass.
O'BRIEN, J. S. (1965). *Science*, **147**, 1099–1107.
OCHS, S. (1966). In *Macromolecules and Behavior*, pp. 20–39, ed. Garto, J. New York: Appleton-Century-Crofts.
PETERS, T., and ASHLEY, C. A. (1967). *J. Cell Biol.*, **33**, 53–60.
RAHMANN, H. (1965). *Z. Zellforsch. mikrosk. Anat.*, **66**, 878–890.
ROBERTSON, J. D. (1962). In *Ultrastructure and Metabolism of the Nervous System*, pp. 94–158, ed. Korey, S. R., Pope, A., and Robins, E. Baltimore: Williams & Wilkins.
SCHMITT, F. O. (1958). *Expl Cell Res.*, Suppl., **5**, 33–57.
SCHMITT, F. O. (1959). In *The Biology of Myelin*, pp. 1–36, ed. Korey, S. R. New York: Hoeber.
SINGER, M. (1952). *Q. Rev. Biol.*, **27**, 169–200.

SINGER, M. (1964). In *Mechanisms of Neural Regeneration*, pp. 228–232, ed. Singer, M., and Schadé, J. P. [*Progress in Brain Research*, vol. 13.] Amsterdam: Elsevier.
SINGER, M. (1965). In *Regeneration in Animals and Related Problems*, pp. 20–32, ed. Kiortsis, V., and Trampusch, H. A. L. Amsterdam: North-Holland Publishing Co.
SINGER, M., and SALPETER, M. M. (1966a). *Anat. Rec.*, **154**, 423 (abstract).
SINGER, M., and SALPETER, M. M. (1966b). *Nature, Lond.*, **210**, 1225–1227.
SINGER, M., and SALPETER, M. M. (1966c). *J. Morph.*, **120**, 281–316.
TAYLOR, A. C., and WEISS, P. (1965). *Proc. natn. Acad. Sci. U.S.A.*, **54**, 1521–1527.
WEISS, P., and PILLAI, A. (1965). *Proc. natn. Acad. Sci. U.S.A.*, **54**, 48–56.

DISCUSSION

Stefanelli: This mechanism may explain certain special conditions seen naturally: for instance, the persistence of a fibre without a cell body, as sometimes happens in very big fish like the tuna when the Mauthner's cells are destroyed but the Mauthner's fibres persist. These fibres without a cell body may be nourished and survive for a long time or persist in a mummified condition without any possible nervous function. To accept the idea of living isolated fibres we need more confirmation of the transport of substances through the myelin. It would be interesting, if it were possible, to change the permeability of Schwann cells by using, for instance, ouabain which blocks the membrane ATPase; if we could then trace a difference in the amino acids within, with or without a change in permeability of the Schwann cells, which are particularly sensitive to ouabain, this would further confirm this mode of transport. In your pictures, Dr. Singer, the labelling is too evenly restricted to the surface; it is not sufficiently concentrated inside the neuron to demonstrate an active process. Experiments on changing this permeability would help to confirm this active transport.

Singer: It is difficult to draw conclusions just from the slides that I have shown. More important are actual counts of silver grains. The latter are most revealing when done on the large fibres of *Xenopus* in which the localization of the label can be better resolved. Our counts, unpublished so far, support the thesis of transport through myelin and Schwann sheath into the axon.

The persistence of fibres separated from their cell bodies is very interesting, Professor Stefanelli, and our attention has also been focused on such cases because axons which receive greater sustenance from their surroundings should, according to our story, persist longer. The survival of such fibres for long periods has been reported for some arthropods. Also, some workers hold the view that in some animals the cut ends of an axon may rejoin without degeneration of the part separated from the cell body. This is reminiscent of what occurs in the single-celled plant, *Acetabularia*, which in a sense is a sort of green neuron because the cell body is located at one end and a long cytoplasmic stalk extends from it. If the cell body is separated from the stalk, the stalk persists and metabolizes for weeks.

Apparently, there is no uniform time of degeneration of transected nerve fibres. In studying the autonomic system of the newt years ago, I was puzzled to find that after four to six weeks these fibres did not degenerate. Perhaps these fibres receive enough from their surroundings to make up for what they have lost from the cell body, at least for a considerable time.

Murray: We observe Schmidt-Lantermann clefts *in vitro* and we have followed the same cleft for a week at a time on the same peripheral nerve. We do not see these clefts in any great numbers in young cultures in which myelin is still being produced and in which the sheath is relatively thin, and is drawn quite tautly over the internode. It is only when the sheaths are 30 to 50 days old that clefts are common. There is also at this period a considerable overgrowth or rippling of the myelin, which is not strictly in the form of clefts. If one observes the clefts in a given internode every day for a week, one finds that some of them come and go. They don't always return to the same position, or return at all. We cannot yet interpret this observation. In the central nervous system where there are very long internodes and some very short ones, do you see any Schmidt-Lantermann clefts?

Singer: Apparently they are there but there are very few obvious clefts in the central nervous system. However, an interruption of the myelin pattern is often seen, sometimes taking the form of bits of oligodendritic cytoplasm interjected in the myelin wrappings. It may be a sort of insensible Schmidt-Lantermann cleft. Or it may be that substances are carried inward only by way of the three-dimensional solid framework of these opposed membranes. There may be forces which operate in a three-dimensional framework to move along substances—forces which we do not understand as yet. It is known that substances can be carried through crystals. In the peripheral system the Schmidt-Lantermann clefts seem to me to be an obvious mechanism of transport. I like to think of them sweeping in a spiral from one end of the internode to the other.

Murray: There might be some other mechanism in the central sheaths. We are trying to take time-lapse pictures of the peripheral sheaths (which are more easily visualized in culture) so that we can record continuously what happens in a single fibre over a period of time.

Hughes: Something of the autonomy of the axon which Professor Singer has described in the adult fibre is seen in its earliest stages of growth. For instance, in cultures of chick dorsal root ganglia one can measure the rate of elongation of the fibres and compare the normal rate with the effect of cutting the fibre. A cut fibre in an explant of chick dorsal root ganglia will go on growing at the normal rate for up to three hours and then it breaks up. But during that three hours its rate of elongation is within the range of the normal. I assume that the explanation is that the fibre elongates primarily from the intake of water along its length, perhaps mostly at the growth terminal, and that growth in the first place is independent of connexions with the cell body.

Singer: Is pinocytosis most evident at the tip of the growing axon? I vaguely remember that W. H. Lewis demonstrated it there.

Murray: Lewis and others saw pinocytotic droplets ascending the axon, but they entered through the growing tip.

DISCUSSION

Gaze: An intimate relationship between the glial cell and the axon is a good thing from my point of view. Have you done a statistical analysis of the distribution of the label over the various structures? Is there in fact a statistically increased probability of getting these grains over the structures you are actually talking about?

Singer: Dr. Salpeter made some counts of the distribution in her electron microscope radioautographs. Others of us have also done counts in the light microscope radioautograph studies. The distribution of grains is not random. The labelling is often heavy. If one accepts the common criteria for labelling by radioautography one sees that all components of the nerve fibres are labelled. We have used five different fixatives: paraformaldehyde before osmium fixation, Bouin's, Carnoy's, formalin, and osmium alone. We have washed before fixation. In all cases a label remains. I am not asserting that the labelled amino acid is incorporated into protein entirely or even in part, although such a conclusion may be justified. There is a possibility that there is some ionic binding of the basic amino acid with the acid phospholipids of the myelin sheath at the time of fixation. We are following up this possibility in our biochemical studies. It may be that the heavy labelling of the sheath is due to a pool of transported free amino acids.

Szentágothai: The evidence about the amino acids seems to be very convincing. But in the case of uridine just from looking at the photographs it seems difficult to tell whether the distribution of the grains is at random, or whether there is a preferential localization over axons or myelin sheaths.

Singer: We do have counts on the uridine label (a paper is now in preparation). The counts are distributed over the various components of the fibres. An important reason why we conclude that uridine, in whatever form, reaches the myelin sheath and axon via the Schwann cell is by comparing these results with those for thymidine. Uridine and thymidine are similar molecules in terms of total charge and biophysical characteristics. Tritiated thymidine hardly labels the nerve, and what does get in is apparently not in the form of DNA because it cannot be removed with DNase. In tritiated uridine the label is removed by procedures commonly accepted as tests for RNA, particularly the use of RNase.

Szentágothai: Where do you suppose that the material moves: in the mesaxon interspace, which is essentially continuous with the extracellular space or, conversely, in the clefts between two Schwann cell membranes, which is continuous with the plasma territory of the Schwann cell, i.e. intracellular? Could clefts be involved?

Singer: We have entertained these possibilities. The space between the faces of the myelin wrappings is of the order of a few Ångströms. Charges on the surface would probably interfere with movement even of a small ion. It is possible that substances may enter at the node of Ranvier and be carried thence along the axon.

Young: I don't think we ought to assume there is protoplasm in the Schmidt-

Lantermann clefts. I went over some photographs with J. D. Robertson recently, but was not convinced that he could find mitochondria in the so-called cytoplasm. I don't know whether they are artifacts or not, but it is not convincingly shown that there is cytoplasm in those clefts.

Singer: It is not convincingly shown that there are mitochondria?

Young: Correct.

Eccles: Do you find any microsomes and so on?

Young: There are no organelles at all.

Singer: My pictures show what we call dark bodies which may be organelles of some interest. We do see many very curious tubules. In well-fixed preparations the tubules may be swollen but seem not to have structured contents. The contents are about as clear as those seen in large areas of astrocytic processes. An organized structured cytoplasm would not be necessary for movement of molecules.

Young: This material is very interesting, but it is not ordinary Schwann cytoplasm.

Eccles: Why did you discount the nodes as a possible factor here?

Singer: We see no special distribution or concentration along the nodes. No special localization occurs which suggests a particular single pathway through or around the sheaths. Consequently, in our theory we tried to involve the whole myelin segment. It is conceivable that labelled substances go in by way of the node but then they would have to spread outwards as well as move into the axon at that point. But we see nothing to suggest such localization, particularly within the first 5 or 10 minutes after injecting the label.

Walton: Do you regard Schwann and Remak cells as being different, and do they behave differently? Is the behaviour different in myelinated as compared to unmyelinated axons?

H. Ford and R. Rhines (1967. *J. neurol. Sci.*, **4**, 501) have used a micro-dissection technique for dissecting out packets of anterior horn cells and central grey and white matter in the rat. They have used a radioactive counting method for working out the uptake of uridine, histidine and lysine in these various packets of cells. This may well be relevant to your work, Dr. Singer.

Singer: I am not familiar with Ford's work. I might remark again that bodies do label with the precursors and often in greater amount than the fibre. And A. Edström (1966. *J. Neurochem.*, **13**, 315–321) has reported the labelling of the axon and myelin by lysine in the Mauthner fibres from segments of the fish spinal cord.

Drachman: Is there other evidence that the amino acid uptake is due to active transport? For example, are some other amino acids *not* transported across the membranes? Do you find that different amino acids of the same type (for example basic, acidic, or neutral) compete with one another for transport? Such competition has been shown in various systems which utilize active transport.

What do you think the nerve cell body normally supplies to its long axoplasmic processes to maintain their integrity? Is it some specialized trophic substance or is it merely additional amino acids and other common nutrients?

Singer: At the moment, we are surveying the amino acids. This is very important because the amino acids about which I spoke, namely histidine and lysine, are highly basic ones. Myelin has many free phosphate groups in it and it may be that non-incorporated basic amino acid is bound ionically to the phosphate groups at the time of fixation. Initial results suggest this may be so because acidic amino acids, for example glutamic and aspartic acid, show much less labelling. There is also less labelling with radioactive proline and threonine. We are combining our survey of the amino acids with biochemical analyses of labelled nerve—this is important because there must be considerable loss of small molecules during fixation. It may be that some other less basic amino acid, for example glutamic acid, also enters the nerve fibre in large quantity but it is not captured by binding during fixation and therefore is washed free.

As for the nature of the trophic contribution from the cell body, there is little that I can add here. The cell body is the trophic centre of the neuron and the axon ordinarily does not survive when separated from it. We know that there is a flow of substance from the cell body into the axon. The important trophic substance may be one or more small molecules, e.g. amino acids, or it may be a large molecule. Our demonstration of another input pathway into the axon may alter our views on the nature of the trophic contribution. It may be that the trophic quality of the axonal substance is fabricated within the axoplasm as well as in the neuronal cell body.

Mugnaini: Have you obtained electron microscope pictures of the uridine incorporation?

Singer: Yes. Our electron microscopy pictures are not unlike those shown for lysine in that there is axon, myelin and Schwann nucleus label. Since osmium fixation prevents later digestion of RNA with ribonuclease, we have not been able to prove that the label is, indeed, RNA. However, in light microscope radioautography we have been able to fix the thin nerves of the newt first in formalin, then to digest the nerves with ribonuclease, and finally to fix a second time in osmium. Our early studies show a reduction in the label after digestion, thus attesting to incorporation of the uridine into RNA. There are reports (e.g. Edström, A. [1966]. *J. Neurochem.*, **13**, 315–321) of the presence of RNA in the myelin sheath and in the axon. Moreover, there are also reports of the presence of enzyme systems in the myelin sheath.

Mugnaini: Are the mitochondria labelled?

Singer: There is label over mitochondria but a statistical study will have to be done to decide whether the mitochondria contain the radioactive material because resolution in electron radioautography is not great. It is not safe to assume a resolution less than 1,000 or even 1,500 Å. An adequate statistical evaluation would require much material and many counts.

TRANSPORT OF MATERIAL ALONG NERVES

G. A. KERKUT

Department of Physiology and Biochemistry, Southampton University

THE central nervous system of the snail (*Helix aspersa*) has certain advantages for physiological research. The ganglia contain large neurons (up to 150 µ in diameter) peripherally arranged so that they are accessible to microelectrodes or injection. Specific neurons can be found from one snail brain to another and thus allow study of the pharmacological and electrophysiological differentiation of the neurons in the central nervous system. The blood system of the snail is mainly haemocoelic and the isolated brain survives well when isolated and placed in Ringer solution; electrical impulses can be obtained from neurons up to 72 hours after isolation.

During a series of experiments on the possible nature of the chemical transmitter at the invertebrate neuromuscular junction, we developed an isolated preparation which consisted of the isolated snail brain, the nerve trunks running from the brain to the pharyngeal muscles, and the pharyngeal muscles. This was a CNS-nerve trunk-muscle preparation and it was possible to stimulate the CNS electrically and obtain a ninhydrin-positive material in the muscle perfusate, this ninhydrin-positive material later being identified as glutamate. The amount of glutamate liberated was proportional to the number of stimuli given to the CNS. Added glutamate would also cause the muscle to contract, and the muscle could convert glutamate to other biochemical compounds (Kerkut *et al.*, 1965; Kerkut, Shapira and Walker, 1965).

The natural corollary of these experiments was to see if they could be repeated with ^{14}C-labelled tracers, and these experiments will be briefly described here. Fuller details are given in Kerkut, Shapira and Walker (1967).

TRANSPORT FROM CNS TO MUSCLE

The snail CNS-nerve trunk-muscle preparation was dissected out and a thick lanolin barrier placed between the brain and the muscle (Fig. 1). The brain rested in a 1 ml. pool of Ringer solution while the muscle was perfused with fresh snail Ringer and the perfusate collected and analysed.

In the first series of experiments the preparation was set up and the brain soaked in 1 μC of uniformly labelled [^{14}C]glutamate for three hours. The muscle was perfused and the perfusate counted in a scintillation counter. There was no significant difference above background count. If the CNS was then electrically stimulated, a radioactive material appeared in the perfusate and the amount of radioactive material was roughly proportional

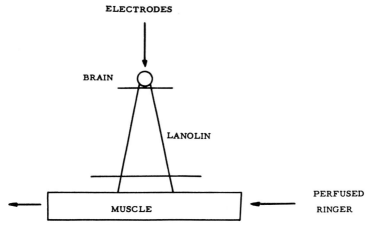

FIG. 1. Diagram of the isolated snail brain–nerve trunk–muscle preparation. The snail brain lies in one compartment. It is separated from the retractor pharyngeal muscle by a thick layer of lanolin. The nerves run through this lanolin. Radioactive material placed in the brain compartment can travel along the nerve trunk and be liberated in the muscle perfusate. (From Kerkut, Shapira and Walker, 1967.)

to the number of stimuli given to the CNS (though prolonged stimulation could exhaust the amount of labelled material). This labelled material when run on thin-layer plates proved to have the same R_F as glutamate in five different solvents. It has therefore been identified as glutamate.

In the next series of experiments the preparation was set up as before and the brain placed in a pool of labelled glutamate. In this case, however, the brain was stimulated immediately it was placed in the glutamate and the muscle perfusate was collected as before. The perfusate showed only background activity for about 20 minutes, after which the amount of radioactivity increased. This labelled material appearing in the perfusate had the same R_F as glutamate and it had taken 20 minutes for the labelled material to travel the 1 cm. of nerve between the CNS and the muscle.

If the brain was soaked in labelled glucose or labelled alanine then the amount of labelled material that appeared in the muscle perfusate was less than when the brain was soaked in glutamate. The material in the perfusate was labelled glutamate even though the brain had been soaked in labelled glucose or labelled alanine. If the brain was soaked in labelled xylose then no labelled material appeared in the perfusate. It would seem that the system is able to transport labelled glutamate along the nerve trunk and that this is specific for glutamate and does not occur for glucose, alanine or aspartate. The conversion takes place in the cell body and not at the muscle endings since the brain can be soaked in labelled glucose, and labelled glutamate can be obtained from the cut end of the nerve trunk.

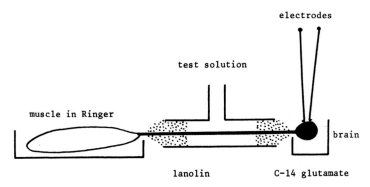

Fig. 2. Diagram of apparatus for studying the effect of cold, Xylocaine or Nembutal on the rate of transport of labelled material from the brain to the muscle. (From Kerkut, Shapira and Walker, 1967.)

EFFECT OF COOLING OR XYLOCAINE ON THE RATE OF TRANSPORT

If the isolated CNS-nerve trunk-muscle preparation is set up as shown in Fig. 2, then one can apply chemicals to the nerve trunk and see how this affects the rate of transport of material from the CNS to the muscle.

Whereas in a "good" preparation it takes about 20 minutes for the labelled glutamate to travel the 1 cm. of nerve from the brain to muscle, when the nerve trunk was cooled to 0°C for the first three minutes of the experiments, or placed in Xylocaine (10^{-4} g./ml.) or Nembutal (10^{-4} g./ml.) for the first three minutes, the liberation of glutamate at the muscle was delayed for more than 30 minutes, i.e. it took more than 50 minutes for the glutamate to appear at the muscle instead of the usual 20 minutes. This effect could not be explained by saying that electrical stimulation was stopped for the three minutes; and the cooling or Xylocaine appear to affect the actual transport of the material along the nerve trunk.

If the snail muscle is incubated in labelled glutamate for 18 to 24 hours, then labelled material appears in the CNS. There is good evidence that labelled material has travelled along the nerve trunk and the rate of transport is slower than that from CNS to muscle. The main worry in all these experiments, and especially those where the system is left for 24 hours, is that there could be a direct leakage from the radioactive pool across the lanolin barrier. For this reason it was decided to try another preparation with a longer piece of nerve trunk. We chose the frog for this.

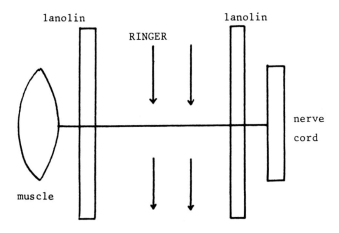

Frog Gastrocnemius--sciatic--nerve cord

FIG. 3. Frog gastrocnemius-sciatic nerve–nerve cord experiment. With a longer nerve trunk it is possible to put a double lanolin barrier between the nerve cord and the muscle and to have Ringer flowing between and so carrying away any material that leaked across the barrier.

When the muscle is soaked in [^{14}C]glutamate for 24 hours, radioactive material is found in the sciatic nerve and nerve cord. (From Kerkut, Shapira and Walker, 1967.)

FROG CNS-SCIATIC NERVE-GASTROCNEMIUS PREPARATION

The vertebral column (and nerve cord) was dissected out together with the sciatic nerve trunk and the gastrocnemius muscle. The experimental set-up was similar to that already used for the snail except that two lanolin barriers were used and a flow of Ringer was kept between the two barriers so as to wash away any labelled material that might have leaked through the lanolin (Fig. 3).

When the nerve cord was soaked in [^{14}C]glutamate solution and electrically stimulated, labelled material later appeared in the muscle perfusate.

This material was glutamate and it took about four hours for the material to travel the 5 cm. of frog sciatic nerve. When the nerve cord was soaked in labelled glucose or alanine and stimulated, the labelled material that appeared in the muscle perfusate was glutamate and not alanine or glucose. The results are thus very similar to those obtained for the snail preparation.

Glutamate has been described coming off from the frog sciatic nerve by Wheeler, Boyarsky and Brooks (1966), there being a release during stimulation of 200 times the resting efflux of glutamate; the release was inhibited by azide but not by ouabain. Stimulation had very little effect on the release

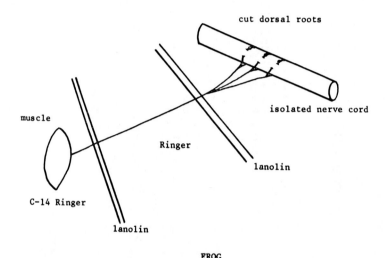

Fig. 4. Effect of cutting the dorsal roots.
Radioactive material still appears in the nerve cord even when the dorsal roots have been cut. This indicates that the material has probably travelled along the motor axons. (From Kerkut, Shapira and Walker, 1967.)

of glycine or aspartate from the frog sciatic nerve, indicating that the release was specific for glutamate. In this case the material was coming from the surface of the nerve trunk. In our experiments we were perfusing the nerve-muscle junctions and the muscle.

If the frog muscle is incubated in labelled glutamate for 24 hours, then labelled material can be found in the nerve cord and in the sciatic nerve. The labelled material still appears in the nerve cord even if the dorsal roots have been cut and plugged with lanolin (Fig. 4). The labelled material would then have travelled antidromically along the motor axons.

The amount of labelled material obtained is not very great but if ten preparations are set up in parallel, with each nerve cord attached to two sciatic nerve trunks and gastrocnemius muscles, it is possible to obtain a greater amount of material from the pooled extracts of sciatic nerve. The material in the sciatic nerve is not glutamate but instead runs in two of the six bands obtained in polyacrylamide electrophoresis (Fig. 5).

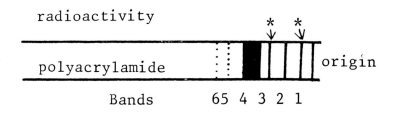

FIG. 5. Diagram of a separation of the labelled material transported from the muscle to the CNS in the sciatic nerve. There are six protein bands, two of which (1 and 3) have labelled material.

TRANSPORT FROM MUSCLE TO NERVE

In those experiments where the gastrocnemius was incubated in labelled glutamate and labelled material was later found in the nerve cord, it was not clear whether the material had been taken up by the nerve endings on the muscle or whether it had been taken up by the muscle and transferred to the nerve. It was in all probability taken up by the nerve endings and not by the muscle. This point is of theoretical importance and the following experiment suggests that material can be carried from muscle to nerve, though it is by no means conclusive.

The sartorius muscle of the frog has its innervation mainly at one end of the muscle. If the sartorius muscle is placed in a polythene tube as shown in Fig. 6 and a clamp placed in the middle of the tube one can prevent the transport of material along the tube.

Labelled material was placed in the nerve-free region of the muscle and the system left for three hours. Labelled material was found in the nerve trunk. This activity was four times higher than the level found in the solution around the nerve, which was just above background.

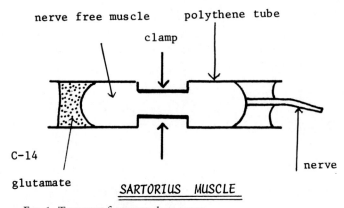

Fig. 6. Transport from muscle to nerve.
If the nerve-free part of the sartorius is soaked in labelled glutamate, labelled material appears in the nerve trunk. The level of activity in the nerve trunk is some four times higher than in the surrounding solution. (From Kerkut, Shapira and Walker, 1967.)

DISCUSSION

The present experiments on the isolated CNS-nerve trunk-muscle preparation indicate that there is a flow of material along the nerve trunk both from cell body to periphery and also from the periphery to the cell body.

There is now a considerable body of evidence to support the view of a flow of material from the cell body to the periphery (Weiss and Hiscoe, 1948; Weiss, 1961; Ochs, Dalrymple and Richards, 1962; Droz and Leblond, 1962; Lubińska, 1964; Miani, 1964; Dahlström and Häggendal, 1966) and the flow rates range from 1 mm./24 hours to 1 cm./hour, depending on the nature of the material transported.

The proteins appear to be transported at a lower rate than, say, the amino storage granules in adrenergic nerves. Our experiments have been concerned with the transport of glutamate and in the snail this is at a very fast rate (1 cm./20 minutes). There appears to be a selective transport of glutamate as compared with glucose, alanine or aspartate. Though there is evidence that glutamate can be liberated from the surface of the frog sciatic nerve during stimulation (Wheeler, Boyarsky and Brooks, 1966) and from crab nerve placed in a sodium-free solution (Baker, 1964), the significance of the glutamate liberation is not yet clear. In those systems where glutamate could be a possible nerve-muscle transmitter, then there is clearly some value in its transport along the axon, though one would have imagined that it would have been easier for it to be synthesized from glucose at the nerve-muscle terminal.

The effect of Xylocaine on the transport of glutamate indicates that the carriage is probably not by simple diffusion, and the fact that glutamate is carried whilst glucose or alanine is not indicates that there is considerable selectivity in the transport system of the neuron.

It is clear that one should not think of the neuron as a bucket with the axon as a hole through which material will just fall. There is an intricate and delicate organization in the transport of material from the cell body to the axon terminals, and as yet we are only just approaching the study of this organization. Tissue cultures of neurons have shown the transport of material both centripetally and centrifugally (Hughes, 1953) but as yet our biochemical understanding of this phenomenon has been limited by lack of experimental procedure. It is hoped that the present system of the isolated CNS-nerve trunk-muscle preparations will help to alleviate this limitation.

Though transport of material from the cell body to the axon terminal is of interest, the transport from periphery to cell body is of even more interest. Such a transport system has been suggested on the basis of studies where axons were transected and changes studied in the cell body (Ochs, Booker and Demeyer, 1961). In some way the cell body "knows" that there has been an interruption in the normal connexions. One would hope that the labelled material that we have found in the sciatic nerve could be part of this message system and work is being carried out in our laboratory to study the variation in the nature of this material with experimental procedures.

If one could theorize from insufficient data, one could suggest that the isolated CNS-nerve trunk-muscle system will allow us to study the relationship between the neurons in the CNS and their peripheral connexions in biochemical terms. If there is a transport of organized material (messenger RNA? proteins for enzyme induction?) across the cell boundary between the nerve and the muscle (Korr, Wilkinson and Chornock, 1967), or from one nerve cell to another, it would allow us to consider the nerve-muscle and the nerve-nerve systems in terms of their chemical differentiation and thus to achieve a new and possibly more fruitful appreciation of metazoan organization in terms of biochemical systems instead of, say, electrical connexions. One could visualize the chemosynthetic systems of the muscle cells being in close harmony with the chemosynthetic systems of the neurons innervating that muscle, with differentiation of the two as a result (Buller, Eccles and Eccles, 1960; Gutmann, 1964). If interneuronal transport was also effective one could consider the close apposition of nerves at synaptic connexion as the result of a similar differentiation.

It is only too easy for theories to outrun facts rapidly, but there is some indication that we can be optimistic and hope that simple isolation preparations and the careful use of tracers and biochemical analysis will allow us to unravel the problems of the organization, growth and maintenance of the nervous system.

SUMMARY

(1) An isolated preparation consisting of CNS-nerve trunk-muscle can be set up with a thick lanolin barrier between the CNS and muscle. The nerve trunk runs through this barrier.

(2) Radioactive material can be placed in the CNS compartment and is found after some time to pass along the nerve trunk and be detected in the muscle compartment.

(3) Radioactive material can be placed in the muscle compartment and after some time will pass along the nerve trunk to the CNS.

(4) Such preparations have been set up for the snail (*Helix aspersa*) brain-nerve trunk-muscle and for the frog nerve cord-sciatic nerve-gastrocnemius muscle.

(5) In the snail [^{14}C]glutamate travels from the CNS to the muscle along 1 cm. of nerve in 20 minutes, It takes 18 to 24 hours for the material to go from the muscle to the CNS.

(6) Transport is slowed by cooling the nerve, by adding Nembutal or Xylocaine to the nerve, or by ensuring that the nerve is not stimulated electrically.

(7) In the frog ^{14}C-labelled material is carried from the muscle to the CNS even if the dorsal roots of the sciatic nerves are cut, thus indicating antidromic carriage along the motor nerves.

(8) Such a compartmental preparation, free from blood circulation, allows one to study possible trophic factors between the CNS and periphery and it also allows the study of the nerve trunk as a chemical system as opposed to the more usual electrical systems.

Acknowledgements

The work described in this paper has been partly financed by grants from the European Research Office of the United States Army, and by the Nuffield Foundation.

REFERENCES

BAKER, P. F. (1964). *Biochim. biophys. Acta*, **88**, 458–460.
BULLER, A. J., ECCLES, J. C., and ECCLES, R. M. (1960). *J. Physiol., Lond.*, **150**, 417–434.
DAHLSTRÖM, A., and HÄGGENDAL, J. (1966). *Acta physiol. scand.*, **67**, 271–277.
DROZ, B., and LEBLOND, C. P. (1962). *Science*, **137**, 1047–1048.
GUTMANN, E. (1964). In *Mechanisms of Neural Regeneration*, pp. 72–112, ed. Singer, M., and Schadé, J. P. [*Progress in Brain Research*, vol. 13.] Amsterdam: Elsevier.

Hughes, A. (1953). *J. Anat.*, **87**, 150–162.
Kerkut, G. A., Leake, L. D., Shapira, A., Cowan, S., and Walker, R. J. (1965). *Comp. Biochem. Physiol.*, **15**, 485–502.
Kerkut, G. A., Shapira, A., and Walker, R. J. (1965). *Comp. Biochem. Physiol.*, **16**, 37–48.
Kerkut, G. A., Shapira, A., and Walker, R. J. (1967). *Comp. Biochem. Physiol.*, **23**, 729–748
Korr, I. M., Wilkinson, P. N., and Chornock, F. W. (1967). *Science*, **155**, 342–345.
Lubińska, L. (1964). In *Mechanisms of Neural Regeneration*, pp. 1–66, ed. Singer, M., and Schadé, J. P. [*Progress in Brain Research*, vol. 13.] Amsterdam: Elsevier.
Miani, N. (1964). In *Mechanisms of Neural Regeneration*, pp. 1–66, ed. Singer, M., and Schadé, J. P. [*Progress in Brain Research*, vol. 13.] Amsterdam: Elsevier.
Ochs, S., Booker, H., and Demeyer, W. E. (1961). *Expl Neurol.*, **3**, 206–208.
Ochs, S., Dalrymple, D., and Richards, G. (1962). *Expl Neurol.*, **5**, 349–363.
Weiss, P. (1961). In *Regional Neurochemistry*, pp. 220–242, ed. Kety, S. S., and Elkes, J. London: Pergamon Press.
Weiss, P., and Hiscoe, H. B. (1948). *J. exp. Zool.*, **107**, 315–395.
Wheeler, D. D., Boyarsky, L. L., and Brooks, W. H. (1966). *J. cell. comp. Physiol.*, **67**, 141–148.

DISCUSSION

Szentágothai: Did stimulation have an effect on the backward transport as well as the forward transport?

Kerkut: We could not find any, but the time variation is very great (18–24 hours in the backward flow); 20–30 minutes for a forward flow is a good experimental time, but for a backward flow 18–24 hours has too big a scatter.

Buller: Have you tried to differentiate between the temperature needed to prevent the transport along the axons and the temperature needed to stop conduction of the electrical impulse, and if so do you get a slowing-down of the transport that is independent of the impulse?

Kerkut: We think that transport is facilitated if impulses are going along, but transport can happen without our applying electrical stimulation. It is very important to choose the right stimulus frequency: if stimulation is at too high a frequency, we cannot pick up any glutamate. The other problem is that we don't know how many nerves in the nerve trunk are electrically active. If stimulation is stopped for a short time, a slight fall-off occurs but it isn't as big as the fall-off one gets by cooling down to 0°C or even 5–6°.

Young: What shocks did you use in these snails at these low frequencies of stimulation?

Kerkut: Square waves of 1 msec. duration for the whole brain, at a low voltage. One mustn't make the muscle twitch; if it does there is less chance of getting the glutamate.

Young: What barrage of impulses do these fibres normally fire? Two or three impulses every three minutes is a pretty low frequency.

Buller: With high frequency stimulation where does the impulse go if it doesn't get down to the end of the axon?

Kerkut: Detecting liberated glutamate is a different matter to detecting action potentials. We are not working on a system like acetycholine where the cholinesterases are inhibited by eserine. If glutamate is the transmitter, we don't know what is destroying the system. With our set-up, by empirical testing, we found the conditions which give the best liberation of glutamate. Many of these neurons are spontaneously active, firing off all the time. When we drive them we do get a greater barrage. The action potentials are long action potentials, lasting perhaps 30–40 msec.

Young: Does this barrage stop the spontaneous discharge?

Kerkut: No.

Eccles: How much goes along if you don't stimulate at all?

Kerkut: If we don't stimulate at all, the material goes along but it takes longer than 20 minutes to get there.

Buller: But there are spontaneous impulses which you can't stop.

Kerkut: We can stop them by cooling the preparation to 0°C.

Eccles: If you cool the brain what happens?

Kerkut: It is then a very long time before we get any labelled glutamate liberated.

Eccles: The effect of cooling on the brain—the emitting side—would be more worth looking at, as regards spontaneous activity, than its effect on the nerve trunk. How does cooling the nerve trunk *per se* have this delaying effect? The nerve trunk has almost no heat capacity, so it must come up immediately to the temperature of the circulating fluid.

Kerkut: One has to distinguish between plasmic flow systems and electrical conducting systems. Low temperatures interfere greatly with this flow system and it takes a long time to recover.

Eccles: Low degrees frighten me. I would like to see you cool it down to perhaps 5° or something like that. I can't understand why it takes such a long time to recover after this three-minute cooling and then when it recovers temperature there is a 25-minute period when nothing happens.

Kerkut: You are quite happy about the action potential recovering quickly but you are not happy about the axoplasmic flow recovering slowly?

Eccles: The low temperature could have caused damage to the whole of the cytoplasmic material in the nerve. I would like to see you investigate different levels of transfer during steady temperatures: to give a shock at low temperatures, let the preparation recover and then watch the reactions is a very mixed experiment.

Kerkut: We chose this time and temperature carefully so that the preparation would recover. What you really want to see is a graph of the rate of flow against temperature.

Gaze: If slowing of transport when fibres are cooled really happens it would strongly suggest to me that the transport mechanism is highly dependent upon the fibre or the Schwann cell rather than upon the cell body, which is not being cooled. Altering the ionic concentration surrounding the fibres could do various funny things. Have you tried altering sodium, potassium, and so forth?

Kerkut: No, we haven't done this. We have been concentrating first of all on establishing the basic parameters, and secondly on trying to analyse the nature of the material that is being transported.

Crain: The maintenance of transport of various important substances down the axon under these conditions of complete block of electrical excitability might have some bearing on the mechanisms that may still continue to operate in the experiments that I described, in which we kept immature brain explants under Xylocaine and other anaesthetics during the entire course of development of synaptic organization in culture. Transport of specific chemicals down the axons of the neurons may play an important role in determining the patterns of neuronal interconnexions that continue to form under conditions of complete depression of bioelectric excitability.

Kerkut: It would be very interesting to increase the concentration of Xylocaine in your preparations to see whether this would ultimately interfere with the establishment of connexions and to find out what the critical concentration is. We were using 10^{-4} g./ml., which is a very high concentration, to block transport as well as electrical impulses.

Crain: We have used Xylocaine in our culture medium at $0 \cdot 5 \times 10^{-4}$ g./ml. for these long-term experiments. With regard to your transport studies, could you comment on the selective extrusion of glutamate from isolated frog sciatic nerve, reported last year by Wheeler and co-workers (Wheeler, D. D., Boyarsky, L. L., and Brooks, W. H. [1966]. *J. cell. comp. Physiol.*, **67**, 141–148)?

Kerkut: We were fortunate to choose glutamate. In the frog we can also put glutamate on the nerve cord and stimulate, and glutamate goes down the trunk quite fast—about 1 cm./hr. or perhaps even slightly faster. This raises a problem, because I don't think that there is very good evidence in the frog that glutamate is the transmitter, unless one wants to suggest a "Burn and Rand" and have glutamate and acetylcholine, like adrenaline and acetylcholine. There is no doubt that glutamate is a special chemical to the CNS. If we put aspartate on the frog nerve cord and stimulate, we get no transport of aspartate. Wheeler and co-workers showed that on stimulation of the nerve trunk there is a 200 per cent increase in the amount of glutamate lost across the axons—across the nerve trunk; this was not inhibited by ouabain but it was inhibited by azide. Aspartate or glycine would not come across. So glutamate is a special compound as far as the nerve is concerned.

Eccles: I am really surprised that glutamate and aspartate differ in this way. When applied electrophoretically to neurons they have a very similar action (Curtis, D. R., and Watkins, J. C. [1965]. *Pharmac. Rev.*, **17**, 347–391). One would like to know to which category some of the other analogues belong.

Kerkut: Of course the inside of the nerve is quite different from the outside.

Hník: Do you find glutamate in the fluid surrounding the muscle, or does some of it really get inside the muscle cell? If so, is there any proof of some circulation inside the muscle cell itself, or does it remain in the vicinity of the end-plate only?

Kerkut: It does get inside. We tested this biochemically by seeing what substances were labelled; we found that glutamate was built into the trichloroacetic acid cycle.

Gutmann: Did you try to block the transport biochemically, for example with actinomycin?

Kerkut: We intend to do this.

Hughes: If you are worried about whether the transport is extra-axonal, wouldn't it be worth while crushing the nerve a few days before you set up the preparation as a control?

Kerkut: It would certainly be a very good idea.

Young: It takes a long time for the frog's nerves to degenerate. You would have to start some months before.

Eccles: Crushing would destroy the axonal continuity. You need a selective interruption of the extra-axonal channels.

Kerkut: That is a negative experiment—to show that something doesn't work. We can get this to happen very easily in good preparations!

Drachman: Have you used tetrodotoxin or saxitoxin?

Kerkut: We have tried tetrodotoxin and it has no or very little effect, but it is a complicated affair (Chamberlain, S.G., and Kerkut, G. A. [1967]. *Nature, Lond.*, **216**, 89).

Mugnaini: Dr. P. R. Flood (1966. *J. comp. Neurol.*, **126**, 181–218; 1967. In preparation) has shown that in *Amphioxus* there is a peculiar mode of innervation. The ventral root fibres appear to be projections from muscle lamellae of the myotome and not nerve fibres. The projections end in conical expansions at the surface of the spinal cord, separated from an extensive layer of axon endings or *boutons* by a thin cleft containing a basement membrane. He calls this system the "central end-plate". Two groups of fibres are present in the ventral roots, one group of thin fibres coming from peripheral lamellae only, and one of thick fibres coming from the rest of the myotome. The two groups of fibres end in face of two separate compartments of *boutons* provided with synaptic vesicles 1,000 Å in diameter in one compartment and 500 Å in the other compartment. Recent studies have shown that the size of the synaptic vesicles in nerve endings related to fast versus slow muscle fibre is different also in other animal species. This could parallel finer biochemical differences. Your preparations, Dr. Kerkut, represent fast muscle fibre systems. It would be interesting to repeat your experiments with a slow system.

DEVELOPMENT AND MAINTENANCE OF NEUROTROPHIC RELATIONS BETWEEN NERVE AND MUSCLE

E. Gutmann

Institute of Physiology, Czechoslovak Academy of Sciences, Prague

Intercellular relations in the neuromuscular system apparently depend on two basic mechanisms: those serving transmission of nerve impulses and excitation of nerve and muscle cells, and those maintaining and restoring the structure and functional capacity of cells, that is, long-term "trophic" regulations between nerve and muscle cells. Though they are of course closely related and connected, the somewhat schematic differentiation of processes of communication and maintenance in intercellular regulations forces us to define more clearly the nature of the "trophic" or maintenance functions in the neuromuscular system. Denervation and reinnervation studies suggest that trophic influences proceed along the nerve cell in both directions to establish a "double dependence" (Young, 1946), and they show that special metabolic connexions are necessary for the maintenance of both neuron and muscle.

After nerve section, the nerve cell increases in volume and contains more proteins (Brattgärd, Edström and Hydén, 1958), and incorporation of [^{35}S]methionine is already increased three days after nerve section (Gutmann et al., 1962). This increased proteosynthesis, demonstrated in radioautographic studies, occurs also after section of the ventral roots (Gutmann, 1964) and is thus independent of sensory nervous activity; it is also marked in dorsal ganglion cells after section of the peripheral cell process (Scott, Gutmann and Horský, 1966), which indicates that this cell reaction too is independent of the direction of the nerve impulse (Fig. 1). This finding underlines the view that the primary trophic role of the perikaryon is related to the outgrowth and maintenance of the axon (Bodian, 1962) rather than to "dynamic polarization", as assumed by Ramón y Cajal (1911). In motor neurons of old animals, the increase in proteosynthesis is still absent even six days after nerve section (Gutmann et al., 1962), and this is apparently connected with the delay in outgrowth of axons which is observed in old animals (Drahota and Gutmann, 1961). Increased proteosynthesis in neurons is thus related to the continuous renewal of proteins, accentuated during the regeneration process. Processes of renewal of

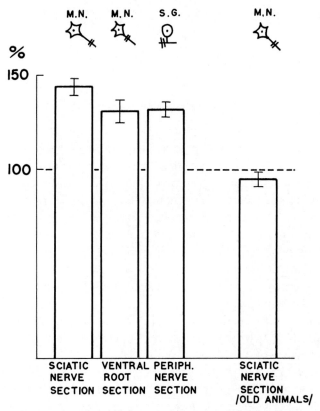

FIG. 1. Percentage change in number of grains in motor neurons (M.N.) and spinal ganglia cells (S.G.) as shown by radioautography three days after sciatic nerve, ventral root and high nerve section in young animals and six days after sciatic nerve section in old animals. (Data from Gutmann, 1964; Scott, Gutmann and Horský, 1966.)

proteins and the axoplasmic flow (Weiss and Hiscoe, 1948; Young, 1945; Lubińska, 1952; Ochs and Johnson, 1967; and others) maintain the neuron, though the physiological significance of the latter is still not clear. Maintenance of the neuron is impaired in old animals and this is demonstrated by the delay in the increase of proteosynthesis after nerve section and by the reduced axoplasmic streaming, demonstrated after nerve crush by lack or delay in the damming up of the enzyme cholinesterase (Fig. 2) (Gutmann and Hanzlíková, 1967b) which normally takes place above and also below (Zelená and Lubińska, 1962) the crush lesion. These few experiments should show a close connexion of the maintenance function of the neuron

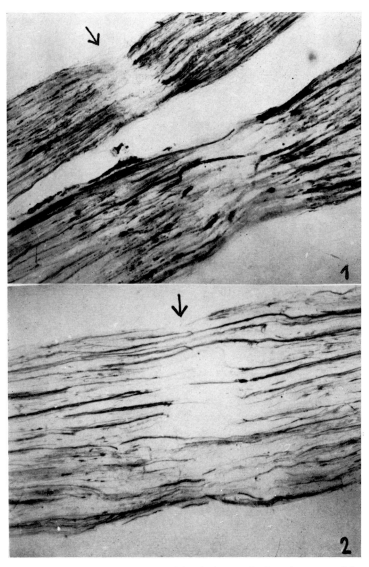

FIG. 2. Cholinesterase activity (acetylthiocholine method) in the axons of the sciatic nerve four hours after crushing the nerve, above and below the crush in young (1) and old (2) animals. Arrows indicate site of nerve crush (Gutmann and Hanzlíková, 1967b).

with the processes of continuous renewal of proteins and transport phenomena.

How do the proteins produced by the neuron participate in muscle maintenance? Denervation studies have repeatedly been used to demonstrate a special neurotrophic function of the nerve cell maintaining the post-synaptic structures. Both denervation and disuse are followed by atrophy, but there are basic differences between these two conditions (see Gutmann, 1962). In denervation, the muscle becomes an unregulated system. Greater and more widespread sensitivity to acetylcholine—normally maintained only at the end-plate region (Ginetzinsky and Shamarina, 1942; Axelsson and Thesleff, 1959; Miledi, 1960)—development of fibrillation activity (Denny-Brown and Pennybacker, 1938), and marked changes of excitability and slowing of the contractile process have been described since Erb (1868) and are apparently related to loss of nerve influence on membrane stability, shown for example by the increase in membrane resistance and reduction of the threshold for excitation (Nicholls, 1956; Jenkinson and Nicholls, 1961). None of these changes can be reproduced by disuse (Tower, 1937; Eccles, 1941; Denny-Brown and Brenner, 1944; and others).

Loss of nerve control, resulting in the muscle becoming an unregulated system, affects both membrane characteristics and intracellular metabolism independent of nerve impulses. A characteristic feature is the increase in deoxyribonucleic acid (DNA) content which also cannot be reproduced by block of neuromuscular transmission (Gutmann and Žák, 1961). Increased incorporation of amino acids into the proteins of denervated muscles has often been reported, though some of the results have not been unequivocal (see Žák, 1962). Increased incorporation occurs only in the sarcoplasmic proteins (Hájek *et al.*, 1966) and the increased proteosynthesis is especially due to an increased supply of substrate, whereas in the reinnervated muscle increased proteosynthesis occurs with normal levels of free amino acids (Hájek *et al.*, 1966). The disturbance of regulation is due especially to a shift of proteosynthesis to other sites, there being apparently a more active synthesis of sarcoplasmic reticulum (Muscatello, Margreth and Aloisi, 1965). This can also be seen in muscles "undergoing denervation hypertrophy", that is in the anterior latissimus dorsi (Feng, Jung and Wu, 1963) and in the diaphragm (Hájek, Haníková and Gutmann, 1967). The denervated hemidiaphragm presents special features, as it is so far the only muscle in which increased synthesis of contractile proteins has been observed (Stewart, 1955; Gutmann *et al.*, 1966). However, in this case there is a decrease of ATPase activity in the contractile proteins, explaining the slowing down of contraction time. To understand the nature of the intracellular disturbance we must therefore look into the control of synthesis of specific proteins. The induction of subneural and myotendinous cholinesterase during development (Zelená and Szentágothai, 1957) and reinnervation by the nerve terminals (e.g. Lubińska and Zelená, 1967) are the best-known examples. Of special interest is the regulation of proteolytic enzymes. The presence of an inhibitory system in the nerve is indicated by changes of proteolytic activity during denervation and reinnervation. During denervation proteolytic activity in muscle increases progressively; during reinnervation normal levels are recovered (Hájek, Gutmann and Syrový, 1964). Increased proteolytic activity after denervation occurs later in a muscle with a long peripheral nerve stump than in a muscle with a short nerve stump. This experiment shows that this change is not linked to loss of nerve impulse activity. It depends apparently on chemical systems which regulate degradation of proteins, are supplied by the nerve and are exhausted earlier in a muscle with a short nerve stump. Experiments in which the motor nerve is severed at different distances from the motor end-plates give some indication of the nature of the maintenance systems supplied by the nerve. For example, degeneration of end-plates and a decrease in glycogen content (Gutmann, Vodička and Zelená, 1955), increase in proteolytic activity (Hájek, Gutmann and Syrový, 1964), decrease in cholinesterase activity (Gutmann and Hanzlíková, 1967b) (Fig. 3), fibrillation and acetylcholine sensitization (Luco and Eyzaguirre, 1955) occur earlier the closer the nerve section is to the end-plate; however the latter changes occur two to three days later than the intracellular changes mentioned.

In this connexion it is of interest that the spread of sensitivity to acetylcholine occurring in denervated muscle indicates that the nerve exerts a direct inhibiting influence on the chemosensitivity of the muscle fibre membrane (Miledi, 1962). Also this neural "inhibitory" factor exerts its control independently of nerve impulses (Miledi, 1960) and it seems unlikely that this controlling "inhibitory" factor is the transmitter itself (see Miledi, 1962). Inhibitory influences by the nerve are also suggested in the control of degradation of proteins. The quantal release of acetylcholine connected with the spontaneous discharge of miniature end-plate potentials (Fatt and Katz, 1952) as the factor controlling sensitivity (Thesleff, 1960) and as a source of the motor nerve's trophic influence (Drachman, 1964) has been assumed. However the sensitivity-controlling agent is apparently not the transmitter substance (Miledi, 1960, 1962) and it is unlikely that trophic control is exerted by acetylcholine only.

Nerve influences affect not only maintenance of structure but also a specific pattern of metabolism in muscles of different function. Two basic types (fast and slow) of muscle and muscle fibre can be distinguished, adapted to requirements of speed and maintenance of tension, and this is coupled to a specific pattern of metabolism. This pattern is established during development and it concerns, for example, a higher rate of proteosynthesis in slow muscles (Gutmann and Syrový, 1967). The importance of the nerve for

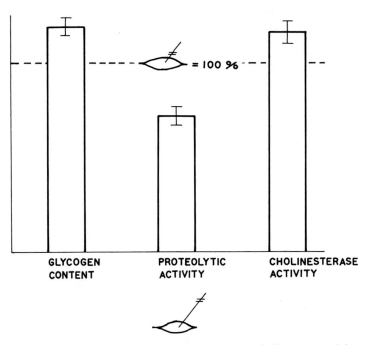

FIG. 3. Glycogen content, proteolytic activity and cholinesterase activity in the extensor digitorum longus muscle of the rat 24 hours after the sciatic nerve had been cut high up in the thigh, compared with the values in the contralateral muscle after the sciatic nerve had been cut near the entry into the muscle (100 per cent).

the maintenance of muscle type differentiation is shown in denervation (Drahota and Gutmann, 1963), cross-union (Drahota and Gutmann, 1963; Guth and Watson, 1967) and heteroinnervation experiments (Gutmann and Hanzlíková, 1967a), and is related to the changes in contraction time (Buller, Eccles and Eccles, 1960) after the nerve supply to fast and slow muscles has been changed. Implantation of a "fast" nerve (the peroneal) into the "slow" soleus muscle at the same time as the tibial nerve innervating the soleus muscle was crushed produced muscle fibres with two types of end-plates, "fast" and "slow" ones (Fig. 4). Such a "fast" additional nerve supply led to an increase in weight, fibre size and total glycogen content and affected also the excitation-contraction coupling, that is, it decreased the contracture response of the soleus muscle to caffeine (Gutmann and Hanzlíkova, 1967a) (Fig. 5). The "fast" peroneal nerve apparently mediates influences restricting the contracture response. The extensor digitorum

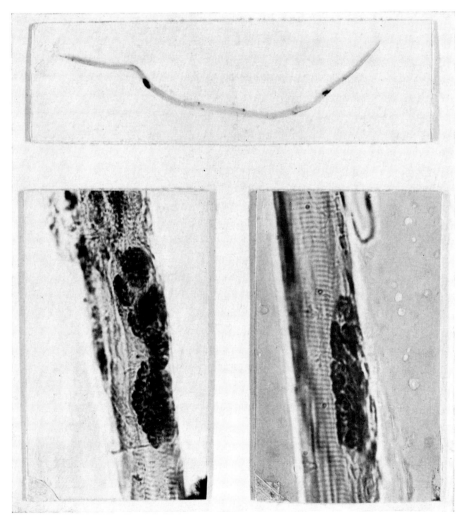

FIG. 4. Single muscle fibre with two end-plates (stained for acetylcholinesterase activity) five months after the tibial nerve was crushed and the deep peroneal nerve simultaneously implanted into the proximal half of the soleus muscle (× 40). The two end-plates of the single muscle fibre, after implantation of the peroneal nerve and after reinnervation of the tibial nerve, are shown below (× 600).

muscle innervated by this nerve normally does not react to caffeine with a contracture (Gutmann and Sandow, 1965). It is improbable that these changes are mediated by processes affecting acetylcholine release as the caffeine contracture can be produced without depolarization of the muscle membrane (see Gutmann and Sandow, 1965).

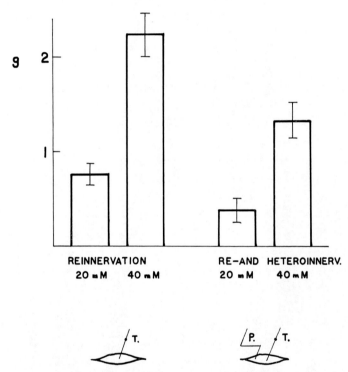

Fig. 5. Caffeine contracture expressed in g. of tension developed isometrically *in vitro* in the soleus muscle six weeks after implantation of the peroneal nerve (P.) and simultaneous crushing of the tibial nerve (T.) on one side, and after crushing of the tibial nerve only on the other side. The values show tension developed by the contracture after addition of 20 and 40 mM solutions of caffeine to the bathing solution.

Thus we must assume that in the nervous control of muscle maintenance not only the mediator, acetylcholine, but probably also many chemical systems supplied by the nerve affect the stability of the membrane and intracellular constituents. Maintenance of muscle is, however, controlled not only by direct nerve influences but also by hormones, which are activated indirectly by the action of nerve cells. The levator ani muscle because of its great sensitivity to hormones, that is testosterone (Wainman and Shipounoff, 1941), is especially suited for such studies. It has been generally assumed that this muscle exists only in the male rat (Hayes, 1965). However, it has been shown that the levator ani develops in both sexes and that from the 18th day of embryological development a progressive involution of this muscle occurs in the female rat, until in the adult female it cannot be found (Čihák, Gutmann and Hanzlíková, 1967). The muscle can, however, persist in female rats if androgens are applied before or at the time of birth, that is, before involution of the muscle blastema (Gutmann, Hanzlíková and Čihák, 1967). In this case a hormone as inducer can maintain the muscle and it can do so also in the absence of innervation, i.e. if the muscle is denervated at birth (Hanzlíková, Gutmann and Čihák, 1967). Electron microscopic studies suggest that the synthesis of proteins induced by this hormone is preferentially shifted to other sites, synthesis of filaments is deficient, and there is a relative increase of sarcoplasmic reticulum and mitochondria. The site of action of androgens is, of course, quite different from that of the "neurotrophic" agent or agents, as can be shown by comparing denervation and castration. Denervation leads to an increase of DNA (Gutmann and Žák, 1961), but castration does not produce this effect (see Venable, 1966); excitability changes (shown by strength-duration curve) and slowing of contraction time are seen very early in the denervated levator ani muscle, but are very small or negligible even in the extremely atrophic levator ani muscle of castrated animals (Hanzlíková, Gutmann and Čihák, 1967).

Androgens modify the concentration or total amounts of many enzymes, but so far there is no indication of an induction effect on synthesis of a specific enzyme protein (see Frieden, 1964) comparable to the acetylcholine-cholinesterase relationship. However, this may be the case, for example, with ecdyson, a hormone controlled by neurosecretory cells. The theory that hormones are general inducers has been put forward by Karlson (1961). It is possible that androgens may stimulate an RNA species with messenger properties.

But even testosterone as a "trophic" hormone and possible "inducer" is under the control of the adenohypophysis via hypothalamic neurons, and affects growth processes in the muscle. A multiple regulation of synthesis of specific proteins establishing and maintaining a specific enzymic outfit in the muscle must be assumed which includes direct and indirect control by the nervous system. The maintenance of neurotrophic relations thus appears as a homeostatic function of the organism closely related to processes of renewal of proteins in the neuron, directly or indirectly affecting the muscle.

SUMMARY

In neurotrophic relations between nerve and muscle cell maintenance of the neuron as well as of the muscle should be distinguished. Experimental evidence shows that the maintenance function of the neuron is closely connected with processes of continuous renewal of proteins and transport phenomena. Nervous influences affect maintenance of structure (i.e. stability of membrane and intracellular constituents) and of a specific pattern of metabolism of muscles of different function. After denervation the muscle becomes an unregulated system, the pattern of metabolism being characterized by increased synthesis of sarcoplasmic proteins and increased activity of proteolytic enzymes, the result being a loss of maintenance and subsequent atrophy. The importance of the disturbance of the control of synthesis of specific proteins is shown. Experimental evidence suggests that acetylcholine cannot be the only source of the trophic influence of the motor nerve. Moreover hormonal influences participate in the maintenance of muscle and, as shown by the example of the effect of testosterone in female rats, they can maintain the muscle even after denervation at birth. The maintenance of neurotrophic relations thus involves a multiple regulation of synthesis of specific proteins and appears thus as a homeostatic function of the organism.

REFERENCES

AXELSSON, J., and THESLEFF, S. (1959). *J. Physiol., Lond.*, **147**, 178.
BODIAN, D. (1962). *Science*, **137**, 323.
BRATTGÄRD, S. O., EDSTRÖM, J. E., and HYDÉN, H. (1958). *Expl Cell Res.*, Suppl., **5**, 185.
BULLER, A. J., ECCLES, J. C., and ECCLES, R. M. (1960). *J. Physiol., Lond.*, **150**, 417.
ČIHÁK, R., GUTMANN, E., and HANZLÍKOVÁ, V. (1967). *Anat. Anz.*, in press.
DENNY-BROWN, D., and BRENNER, C. (1944). *J. Neurol. Neurosurg. Psychiat.*, **7**, 76.
DENNY-BROWN, D., and PENNYBACKER, J. B. (1938). *Brain*, **61**, 311.
DRACHMAN, D. B. (1964). *Science*, **145**, 719.
DRAHOTA, Z., and GUTMANN, E. (1961). *Gerontologia*, **6**, 181.
DRAHOTA, Z., and GUTMANN, E. (1963). *Physiologia bohemoslov.*, **12**, 339.
ECCLES, J. C. (1941). *Med. J. Aust.*, **2**, 160.
ERB, W. (1868). *Dt. Arch. klin. Med.*, **5**, 42.
FATT, P., and KATZ, B. (1952). *J. Physiol., Lond.*, **117**, 109.
FENG, T. P., JUNG, H. W., and WU, W. J. (1963). In *The Effect of Use and Disuse on Neuromuscular Functions*, pp. 432–441, ed. Gutmann, E., and Hník, P. Prague: Czechoslovak Academy of Sciences.
FRIEDEN, E. H. (1964). In *Actions of Hormones on Molecular Processes*, pp. 509–559, ed. Litwack, G., and Kritchevsky, D. New York: Wiley.
GINETZINSKY, A. G., and SHAMARINA, N. M. (1942). *Usp. sovrem. Biol.*, **15**, 283.
GUTH, L., and WATSON, P. K. (1967). *Expl Neurol.*, **13**, 51.
GUTMANN, E. (1962). *Revue can. Biol.*, **21**, 353.
GUTMANN, E. (1964). In *Mechanisms of Neural Regeneration*, pp. 72–114, ed. Singer, M., and Schadé, J. P. [*Progress in Brain Research*, vol. 13.] Amsterdam: Elsevier.

GUTMANN, E., HANÍKOVÁ, M., HÁJEK, I., KLICPERA, M., and SYROVÝ, I. (1966). *Physiologia bohemoslov.*, **15**, 508.
GUTMANN, E., and HANZLÍKOVÁ, V. (1967a). *Physiologia bohemoslov.*, **16**, 244.
GUTMANN, E., and HANZLÍKOVÁ, V. (1967b). Unpublished observations.
GUTMANN, E., HANZLÍKOVÁ, V., and ČIHÁK, R. (1967). *Experientia*, **23**, 852
GUTMANN, E., JAKOUBEK, B., HÁJEK, I., and ROHLÍČEK, V., and ŠKALOUD, J. (1962). *Physiologia bohemoslov.*, **11**, 437.
GUTMANN, E., and SANDOW, A. (1965). *Life Sci.*, **4**, 1149.
GUTMANN, E., and SYROVÝ, I. (1967). *Physiologia bohemoslov.*, **16**, 232.
GUTMANN, E., VODIČKA, Z., and ZELENÁ, J. (1955). *Physiologia bohemoslov.*, **4**, 200.
GUTMANN, E., and ŽÁK, R. (1961). *Physiologia bohemoslov.*, **10**, 493.
HÁJEK, I., GUTMANN, E., KLICPERA, M., and SYROVÝ, I. (1966). *Physiologia bohemoslov.* **15**, 148.
HÁJEK, I., GUTMANN, E., and SYROVÝ, I. (1964). *Physiologia bohemoslov.*, **13**, 32.
HÁJEK, I., HANÍKOVÁ, M., and GUTMANN, E. (1967). *Physiologia bohemoslov.*, **16**, in press.
HANZLÍKOVÁ, V., GUTMANN, E., and ČIHÁK, R. (1967). *Čslká Fysiol.*, **16**, in press.
HAYES, K. J. (1965). *Acta endocr.*, *Copenh.*, **48**, 337.
JENKINSON D. H., and NICHOLLS, J. G. (1961). *J. Physiol., Lond.*, **159**, 111-127.
KARLSON, P. (1961). *Dt. med. Wschr.*, **86**, 668.
LUBIŃSKA, L. (1952). *Acta Biol. exp., Lodz*, **16**, 73.
LUBIŃSKA, L., and ZELENÁ, J. (1967). *J. Anat.*, **101**, 295.
LUCO, J. V., and EYZAGUIRRE, C. (1955). *J. Neurophysiol.*, **18**, 65.
MILEDI, R. (1960). *J. Physiol., Lond.*, **151**, 1.
MILEDI, R. (1962). *Ciba Fdn Symp. Enzymes and Drug Action*, pp. 220-235. London: Churchill.
MUSCATELLO, U., MARGRETH, A., and ALOISI, M. (1965). *J. Cell Biol.*, **27**, 1.
NICHOLLS, J. G. (1956). *J. Physiol., Lond.*, **131**, 1.
OCHS, S., and JOHNSON, M. H. (1967). *J. Neurochem.*, **14**, 317.
RAMÓN Y CAJAL, S. (1911). *Histologie du Système Nerveux*. Paris: Maloine.
SCOTT, D., GUTMANN, E., and HORSKÝ, P. (1966). *Science*, **152**, 787.
STEWART, D. M. (1955). *Biochem. J.*, **59**, 553.
THESLEFF, S. (1960). *Physiol. Rev.*, **40**, 734.
TOWER, S. S. (1937). *J. comp. Neurol.*, **67**, 241.
VENABLE, J. H. (1966). *Am. J. Anat.*, **119**, 263.
WAINMAN, P., and SHIPOUNOFF, G. C. (1941). *Endocrinology*, **29**, 975.
WEISS, P., and HISCOE, A. B. (1948). *J. exp. Zool.*, **107**, 315.
YOUNG, J. Z. (1945). In *Essays on Growth and Form*, pp. 41-94, ed. Le Gros Clark, W. and Medawar, P. B. Oxford University Press.
YOUNG, J. Z. (1946). *Lancet*, **2**, 109.
ŽÁK, R. (1962). In *The Denervated Muscle*, pp. 273-340, ed. Gutmann, E. Prague: Czechoslovak Academy of Sciences.
ZELENÁ, J., and LUBIŃSKA, L. (1962). *Physiologia bohemoslov.*, **11**, 261.
ZELENÁ, J., and SZENTÁGOTHAI, J. (1957). *Acta histochem.*, **3**, 284.

DISCUSSION

Hník: May I introduce briefly a new aspect into these neurotrophic relations—an aspect which may ultimately prove useful in understanding the way in which trophic processes are being integrated, that is, a reflexogenic mechanism.

This concerns the increased sensory outflow from muscles undergoing atrophy. The original finding of W. Kozak and R. A. Westerman (1961. *Nature*,

Lond., **189**, 753) was further studied in our laboratory and it became evident that this increased sensory activity was not apparently due to discharges in proprioceptive nerve fibres, since no difference was found in their mean frequency of discharge as compared with control muscles (Hník, P., Beránek, R., Vyklický, L., and Zelená, J. [1963]. *Physiologia bohemoslov.*, **12**, 23).

At this time we therefore put forward the hypothesis that perhaps it is the non-proprioceptive nerve endings which might be responsible for this activity. If this were actually so and the free sensory nerve endings—which represent a quarter to a third of all myelinated sensory nerve fibres in a muscle nerve (Barker, D., Ip, M. C., and Adal, M. N. [1962]. In *Symposium on Muscle Receptors*, p. 257, ed. Barker, D. Hong Kong University Press; Zelená, J., and Hník, P. [1963]. *Physiologia bohemoslov.*, **12**, 277)—were being stimulated by some change or changes occurring in the muscle in the course of atrophy, then other types of atrophy should also be accompanied by increased sensory outflow.

This was actually found to be the case, since chronic ventral root section caused a 70 per cent increase in sensory outflow, as measured in the peripheral nerve (Hník, P. [1964]. *Physiologia bohemoslov.*, **13**, 405). When dorsal root filaments were teased and the population of spontaneously firing nerve endings was assessed, it was found that while in normal muscles only 3 per cent of spontaneously discharging endings are of non-proprioceptor origin, they represent almost a half (45 per cent) of fibres with spontaneous activity in chronically de-efferented muscles, which in control muscles are silent. This could explain why the sensory outflow from de-efferented muscles is increased (Hník, P., and Payne, R. [1966]. *Physiologia bohemoslov.*, **15**, 498).

These findings have, apparently, a more general pathophysiological implication. Several pathological clinical syndromes, such as gastric ulcers, reflex osteoporosis, trophic ulcers and others have been hypothetically explained by a vicious circle mechanism. However, no concrete data have, as yet, been available about the sensory output from pathological sites. In the experiments referred to above, increased sensory outflow has been demonstrated from a pathologically changed organ, i.e. atrophying muscle, and a certain indication has also been obtained that this sensory outflow adversely affects the pathological process which initiated it. Interruption of this continuous flow of sensory activity by dorsal root section has been shown to retard the rate of atrophy very substantially (Hník, P. [1964]. *Physiologia bohemoslov.*, **13**, 209). The long-lasting increase in sensory outflow from the muscle evidently affects synaptic transmission at the spinal level, since the monosynaptic response after tenotomy is considerably enhanced (Beránek, R., and Hník, P. G. [1959]. *Science*, **130**, 981; Kozak, W., and Westerman, R. A. [1961]. *Nature, Lond.*, **189**, 753; Beránek, R., *et al.* [1961]. *Physiologia bohemoslov.*, **10**, 543).

All these findings seem to indicate that a pathological state may give rise to sensory nervous activity, which may affect both central and peripheral processes.

One further point concerns hormonal and neural effects in neurotrophic relations. It is interesting to realize that hormonal effects apparently play a much greater role at early developmental stages than later in life. Dr. Zelená and I (1957. *Physiologia bohemoslov.*, **6**, 193) found that for about the first 14 days after denervation at birth, rat muscle fibre diameters still go on increasing, as does muscle weight. Growth proceeds more slowly than in control muscles but it is still quite contrary to the rapid denervation atrophy which occurs later in life. This finding was first made by G. Schapira, J. C. Dreyfus and F. Schapira (1950. *C.r. Séanc. Soc. Biol.*, **144**, 829) in rabbits, which showed an increase in weight and protein content after denervation during the early postnatal period. It would thus appear that during early developmental stages, there are evidently hormonal influences which co-operate with the neural mechanisms in maintaining the muscle and ensuring its growth.

Walton: Some of your work was done on the rabbit, Dr. Gutmann, but there is something very peculiar about rabbit muscles, both metabolically and electrically. In work on chloroquine-induced myopathy in the rabbit (to be published shortly), my colleagues Drs. A. de Aguayo and P. Hudgson have found that the creatine kinase activity in the serum is extraordinarily labile. Simply handling a rabbit subjects it to psychic trauma which can produce a tenfold increase in serum creatine kinase within a very short period.

There is also something very peculiar about the muscle fibre membrane in rabbits, because electromyographically one can record a great deal of pseudomyotonic activity in the apparently normal rabbit which we have been quite unable to explain. This kind of instability of the membrane is not seen in the rat, the mouse or the guinea pig. Work carried out on rabbit muscle from a metabolic standpoint should therefore be interpreted very cautiously.

Gutmann: Most of the experiments reported here were performed on rats. There are certainly species differences concerning the metabolism of muscles but the general pattern is very similar. An interesting difference is that after denervation the glycogen content is increased in rabbit muscles for a long time but only for a very short time in rat muscles. However, at later stages there is a decrease of glycogen in the denervated muscles of both species (Gutmann, E. [1959]. *Am. J. phys. Med.*, **38**, 104).

Szentágothai: You mentioned that when you cut the axon distal to the spinal ganglion the effect is much stronger than if you cut the central process. But then you say also that it has nothing to do with the direction of impulse conduction. How do you arrive at this conclusion?

Gutmann: I was referring to the experiments in which the peripheral cell process is cut. In this case there is an increase of proteosynthesis. If the central cell process is cut no increase could be found (Scott, D., Gutmann, E., and Horský, P. [1966]. *Science*, **152**, 787). I think this is because the fibres are thinner, the related loss of proteins after section being smaller. An adaptive reaction may also be included, the central fibres entering the cord where no regeneration occurs.

Szentágothai: There is not very much chromatolysis in the spinal ganglion cell if the dorsal root is cut, whereas there is very strong chromatolysis if you cut the sensory nerve. If the dorsal root is cut in the young animal no reduction of fibre diameter and myelin sheath thickness is experienced after six weeks or so in the sensory nerve fibres distal to the ganglion. A very strong reduction in both is found in dorsal root fibres if under otherwise identical circumstances the cut has been made distal to the ganglion. As a similar reduction of fibre diameter and myelin sheath thickness could be achieved in sensory fibres by removing the long bones from the hindleg in the young dog—i.e. by removing a considerable part of sensory stimulation—we interpreted these effects as due to the lack of impulse flow (Szentágothai, J., and Rajkovits, K. [1953]. *Acta morph. hung.*, **5**, 253-274).

Gutmann: My interpretation is that the main factor is the degree of loss of proteins after nerve section. It has been pointed out that almost no chromatolysis is seen when the nerve is cut far away from the nerve cell (Geist, F. D. [1933]. *Archs Neurol. Psychiat., Chicago,* **29**, 88-103). In this case there is little loss of proteins after nerve section. The closer the nerve section to the nerve cell, the higher the degree of chromatolysis. This is actually related to the reaction of increased proteosynthesis. I don't think we should use the term "retrograde degeneration", because we actually deal with a progressive reaction of the nerve cell.

Szentágothai: In transneuronal atrophy, induced in very young animals, the effect also spreads much more in the direction of impulse conduction of the neuron chain. In the reverse direction transneuronal atrophy is negligible in most cases. This would also point towards the significance of impulse conduction.

Drachman: Do you think that the increase in DNA production and in protein synthesis may represent attempts at compensation by the denervated muscles? Is it possible that the increase in afferent impulses from tenotomized muscle (shown by Dr. Hník) may likewise represent attempted compensatory activity? This suggests a teleological use for the changes in metabolic and physiological activity, and we shall probably not be able to resolve the point.

Gutmann: I agree.

Hník: It is difficult to say whether this sensory outflow is a sort of SOS which the muscle is sending out while undergoing atrophy. It does not appear to be a favourable mechanism from the point of view of the muscle, since elimination of this sensory outflow by dorsal root section slows down atrophy after tenotomy.

Eccles: I am very worried about the receptors you talked about, Dr. Hník. Surely you don't think that there are many group I receptors that are not in muscle spindles or tendon organs?

Hník: We measured conduction velocities of all these fibres, and some group I fibres not responding to stretch are certainly involved. I was very worried about this too, until I realized that the histological studies of D. Barker, M. C. Ip and M. N. Adal (1962. In *Symposium on Muscle Receptors,* pp. 257-261, ed. Barker, D. Hong Kong University Press) and the work of A. S. Paintal (1961. *J. Physiol., Lond.,* **156**, 498-574) both allow up to 10 per cent of surplus group I fibres, not allotted to proprioceptive endings. There is in fact reasonably good evidence that they are not proprioceptive endings which have lost the ability to respond to stretch as a result of long-term de-efferentation.

Buller: Do you mean they respond to stretch?

Hník: No. Chronically de-efferented proprioceptive endings exhibit increased sensitivity to muscle stretch (Hník, P. [1964]. *Physiologia bohemoslov.*, **13**, 405). But some spontaneously discharging endings do not and we consider these to be non-proprioceptive in origin. It is very unlikely that some proprioceptive endings would have increased sensitivity to stretch while others lose it altogether.

Murray: How does the nerve cell soma know whether it should or should not chromatolyse, when it has a long or a short axon meeting an abnormal situation?

Gutmann: I assume that it is the loss of proteins which gives the main signal.

Murray: The loss of proteins at 5 inches or 10 inches away?

Gutmann: Transport phenomena, the axoplasmic flow, may be involved, but this is still a matter for speculation.

Murray: Is there any retrograde transport of signal substances?

Gutmann: There is a bidirectional flow which may have a signalizing function.

Young: The point is not how it operates, but the fact that the difference exists. We must find a technique with which we can explore what the signalling system may be.

Eccles: This is like your flow story again, Professor Young. There is a lot of resistance to transport down the axon, hence the delays in operation at a distance.

Young: How easily can we be sure that this difference exists, so that we can do more precise studies? You say the nerve endings degenerate in 24 hours with a low cut, and in 48 hours with a high cut, Professor Gutmann. Can you distinguish a degenerating from a non-degenerating end-plate?

Gutmann: There is a breakdown of axonal terminations at the "degenerating" end-plates. One can easily see that in the muscle with the long nerve stump there are practically no degenerated end-plates at a time when "degenerated" end-plates are found in the muscle with the short nerve stump.

Young: That would give us the difference between 24 and 48 hours, but how precise can this measure be? If we could have an accurate measure of the time at which degeneration occurs we could perhaps learn something about how the nerve cell understands when chromatolysis should take place.

Gutmann: I think it can be done, but it would be easier to do it with the help of biochemical criteria.

Levi-Montalcini: In the experiments on nerve transplantation into the diaphragm, did two motor end-plates form on the muscle fibre receiving the additional nerve? If so, how long did the extra motor end-plate persist?

Gutmann: The denervated hemidiaphragm is apparently passively stretched by the contractions of the intact hemidiaphragm. There is an increase of contractile proteins, but they are deficient in their enzymic outfit, having for example a low

ATPase activity. This results in a decreased speed of contraction. The end-plates persist even eight months after implantation and there was no rejection.

Buller: How complete were your double innervations? I have never had more than a few fibres doubly innervated. Have you done your biochemistry only on those fibres which were known to be doubly innervated or on the complete muscle?

Gutmann: Only a relatively small percentage of muscle fibres with two end-plates can be obtained. If the nerve is implanted distally there are only 10 to 15 per cent more end-plates than in the control muscle. A higher number of end-plates is achieved by a more central implantation of the nerve. However, the percentage of doubly innervated muscle fibres is still relatively small.

Buller: If you have a relatively small proportion of doubly innervated fibres, mixed with other fibres receiving one or other innervation, how are you able to interpret your biochemical studies? Did you study tension overlaps?

Gutmann: No, not really. The weakness is that we do not know exactly how many muscle fibres have two end-plates. We would have to tease all the muscle fibres in the muscles and this is very difficult.

Eccles: There are certain muscles in which almost every muscle fibre is doubly innervated, such as the frog sartorius, and I think the cat gracilis.

Buller: But not by heterogeneous motor neurons.

Eccles: No, but this could be arranged.

Buller: I am not sure that it can. I think this is a key point. The mammalian muscle fibre normally takes one innervation, and one has to use rather subtle tricks to get even a minimal amount of double innervation.

Eccles: I would suggest utilizing muscle fibres so long that one end doesn't know what the other end is doing!

Drachman: Professor Gutmann, you mentioned a variety of changes in skeletal muscle which occurred more rapidly with section of the nerves closer to the muscle. Recently, C. R. Slater (1966. *Nature, Lond.,* **209**, 305–306) has studied the rate of disappearance of miniature end-plate potentials after section of nerves at different lengths. He found that their rate of disappearance depended on the length of the remaining nerve stump. Each centimetre of peripheral nerve stump allowed miniature end-plate potentials to continue for about 40 minutes in the rat. This is an important observation, which suggests to me that the other phenomena you have mentioned may follow as a consequence of the loss of spontaneous acetylcholine release. This is the explanation which N. Emmelin and L. Malm favour in their similar studies (1965. *Q. Jl exp. Physiol.,* **50**, 142–145).

Gutmann: The first findings were those on degeneration of end-plates and changes of glycogen content (Gutmann, E., Vodička, Z., and Zelená, J. [1955]. *Physiologia bohemoslov.,* **4**, 200) and on changes of fibrillation activity (Luco, J. V., and Eyzaguirre, C. [1955]. *J. Neurophysiol.,* **18**, 65).

Szentágothai: There is electron microscopic evidence that the length of the stump is important also in secondary degeneration. If the dorsal roots are cut the degeneration in the segment of entrance is extremely quick, whereas with most other longer pathways (optic fibres, spinocerebellar tract) there is a difference of one or two days.

Eccles: You meant the central roots, didn't you?

Szentágothai: No, dorsal roots, and looking at the secondary degeneration of collaterals in the grey matter. The length of the peripheral stump is considered, i.e. how far the cut is from the ending one is looking at. With the electron microscope much more exact timing of the degeneration is possible than with silver stains.

Eccles: Are you cutting the dorsal roots central to the ganglion, but nearer or farther from the entrance of the spinal cord?

Szentágothai: We always cut it quite close to the entrance of the spinal cord. I agree that what you ask would be the correct way to test this, but we have not done this experiment as yet. Probably other pathways would be more suited anatomically for such an experiment. So far this is only a chance observation on various kinds of fibres with various lengths of the peripheral stumps. The speed of degeneration of terminals appears to depend strongly on the lengths of the stump.

Prestige: I am slightly worried about the suggestion that chromatolysis is initiated by disturbances in protein synthesis. Couldn't there be a substance normally coming up fibres from the periphery, in the absence of which chromatolysis occurs? If the cut is far distal, there is more fibre attached, so there is more substance in the axoplasm still to come and the cell would have more time to adjust. J. C. Hinsey, M. A. Krupp and W. T. Lhamon (1937. *J. comp. Neurol.*, **67**, 205–214) cut dorsal roots in the sacral region and found that unless they cut very close to the ganglion and actually got local trauma, they got no loss of cells even two to three years later. Apparently in the sacral region the cord is far enough away from the ganglia and there is no maintenance function from the spinal cord back into the ganglia. However, cutting peripheral to the ganglia, S. W. Ranson (1906. *J. comp. Neurol.*, **16**, 265–293) found extensive loss of cells.

Gutmann: Did he cut the dorsal roots?

Prestige: After dorsal roots were cut there was no loss of cells in the dorsal root ganglion. After peripheral nerves or spinal nerves were cut, most of the cells degenerated. I think this is very similar to the chromatolysis situation.

Gutmann: Communication within the neuron is becoming more complicated. Apparently in the morphogenetic processes other signals are used than in nerve impulse transmission, but their nature has still to be determined.

Eccles: The axon reaction or chromatolysis is probably caused by the drainage down the axon as a consequence of the intensive peripheral regenerative process. When you cut axons centrally to the dorsal root ganglia regeneration is a failure: it is not guided, and it gives up quite quickly, so there is very little metabolic drain on dorsal root ganglion cell.

Prestige: I would question even the drainage idea. There is a big pile-up of axoplasm inside the fibre, but this is a local event; it only happens within a millimetre or two millimetres.

Eccles: But I am thinking of the drainage due to the outgrowing regenerating branches.

Prestige: The disturbances are very local. It might be easy to understand how mechanical events cause the traumatic degeneration at close range, but not at long distances.

Eccles: This is not mechanical. What we are talking about is that the outgrowth of regenerating fibres puts a tremendous protein drain on the protein-synthesizing mechanisms, which are in the cell body.

Prestige: But how does the information get up in the first place?

Eccles: It is not a mechanical process. It is not as if the axon was a pipe filled by merely passive contents. Ciné-photography shows that there is movement of particles up and down at quite a considerable velocity.

Kerkut: And different things go at different speeds.

Murray: There is a constant rate of proximo-distal flow, anyway. What is going to change this through a long, narrow tube with uniform bore? It could hardly go any faster than it does.

Eccles: It can go both ways, and possibly neurotubules contribute to this.

Hughes: One approach to this chromatolysis question should be a comparative one. It is much more difficult to cause chromatolysis in some animals than in others. George Romanes, for instance, in his work on muscular nerves in the cat (1951. *J. comp. Neurol.*, **94**, 313–363) had to crush twice before he could get a response. Then there are the differences between cells of different sizes.

Eccles: And of course the neurons of a kitten exhibit much more chromatolytic change than those of an old cat.

THE ROLE OF ACETYLCHOLINE AS A TROPHIC NEURO-MUSCULAR TRANSMITTER†

Daniel B. Drachman

Laboratory of Neuroembryology, Department of Neurology, Tufts-New England Medical Center, Boston, Massachusetts

The immediate effect of denervation of skeletal muscle is motor paralysis. Over the longer term, denervation leads to a series of alterations in the morphology, physiology and metabolism of muscle which have been termed "atrophic" (Gutmann and Hník, 1962; Gutmann, 1964). It is clear that the motor nerves alone supply the "trophic influence" which is capable of preventing these changes, while the sensory and sympathetic innervations play no significant role (Tower, 1931a, b; 1935). The question of *how* the nerves exert this effect remains of paramount importance.

The weight of present evidence suggests that neither conduction of motor nerve impulses nor mechanical work of muscle are the factors necessary to prevent denervation atrophy of skeletal muscle. Elimination of conducted nerve impulses by various experimental techniques (Tower, 1937; Denny-Brown and Brenner, 1944; Gutmann and Žák, 1961) fails to reproduce fully the effects of denervation, provided that the motor nerves remain anatomically and functionally connected to the muscle. Furthermore, relieving skeletal muscle of its work load (for example, by tenotomy) is not equivalent in its results to denervation. Finally, the atrophy of a denervated muscle cannot be prevented (although it may be retarded) by direct electrical stimulation (Gutmann and Guttmann, 1942; Eccles, 1943).

Alternatively, it has been suggested that acetylcholine (ACh) released at the motor nerve endings may function as the transmitter of the motor nerve's trophic influence (Thesleff, 1960, 1961; Josefsson and Thesleff, 1961; Drachman, 1964). It is unlikely that the ACh released in response to nerve *impulses* is essential for this effect, since impulses can be eliminated (as mentioned above) without producing denervation atrophy. However, it is well established that ACh is also liberated in a manner which is independent of neural impulses. This spontaneous, continual discharge of quanta

† Supported by NIH Grant No. HD 01083.

of ACh is reflected in the miniature end-plate potentials described by Fatt and Katz (1952). It is possible that ACh supplied in this fashion may be adequate to maintain the nerve's trophic influence.

In order to investigate the trophic action of ACh, one may block its production or transmission pharmacologically, without otherwise disturbing the neuromuscular connexions. For this purpose, three pharmacological agents which are known to interfere with cholinergic transmission at different levels have proved to be useful tools: botulinum toxin, hemicholinium no. 3 (HC-3) and curare. Botulinum toxin acts presynaptically, abolishing release of ACh at cholinergic nerve endings (Burgen, Dickens and Zatman, 1949; Brooks, 1956; Lamanna, 1959). HC-3 exerts both pre- and post-synaptic effects. It interferes with synthesis of ACh at cholinergic nerve endings, by preventing uptake of choline. In high concentration, HC-3 also acts post-synaptically, depressing sensitivity of the receptor membrane to ACh (Hofmann, 1966). The principal action of curare is to block access to ACh at the level of the muscle membrane receptor (Taylor, 1959), although a pre-synaptic effect has also recently been observed (Standaert, 1964).

Thesleff and his associates have shown that local treatment of muscle with minute amounts of botulinum toxin reproduces some of the physiological effects of denervation. Thus, fibrillations occur within five to six days in toxin-treated muscle (Josefsson and Thesleff, 1961). Microelectrode studies have revealed that the ACh-sensitive area on the surface of the muscle fibre spreads after botulinum treatment, as it does after denervation (Thesleff, 1960, 1961).

The present investigation was designed to determine whether prolonged treatment with neuromuscular blocking agents would reproduce the *morphological* effects of denervation of skeletal muscle. Study of this question requires an experimental animal able to survive intoxication thorough enough to inhibit neuromuscular transmission completely, and prolonged enough to permit the histological changes of denervation atrophy to take place. For this purpose, the chick embryo in-the-egg is a useful test system. Unlike most adult animals, it readily survives total neuromuscular paralysis (Drachman, 1964), since its respiration does not require muscular activity (Romanoff, 1960).

The chick embryo is an unfamiliar subject for neuromuscular experimentation in two respects. First, it is a bird, rather than one of the more commonly used laboratory mammals. Second, the rapidity of its embryonic growth and differentiation introduce special considerations. However, neither of these features impairs its usefulness. Denervation atrophy in *birds* is known to follow a course similar to that in mammals (Knoll and Hauer, 1892). The fact that the developing embryo is undergoing continuous change may be taken into account by: (1) using age-matched controls in all cases, and (2) ascertaining the stages of normal development of the neuromuscular system, as detailed below:

(1) Total incubation time — 21 days
(2) Earliest movements (Levi-Montalcini and Visintini, 1938*a*) — $3\frac{1}{2}$ days
(3) Innervation of muscle (earliest contact) (Levi-Montalcini and Visintini, 1938*b*) — 4–5 days
(4) Intimate neuromuscular contact (silver stain) (Tello, 1917; Mumenthaler and Engel, 1961) — 7 days
(5) Cholinesterase at end-plates (Mumenthaler and Engel, 1961) — 12–14 days
(6) Active respiratory movements required (Romanoff, 1960) — 20–21 days

Since the present study was designed to explore the interrelation between nerves and muscles *after* the establishment of neuromuscular contact, the seventh day of incubation was chosen as the *earliest* starting point for most experiments. Experiments were also begun on the 12th day, a time when cholinesterase (ChE) accumulation indicates further maturation of the motor end-plates. All series were terminated before the 20th day of incubation, because of the embryo's requirement for active respiration by that time.

In the present context, the chick embryo is regarded both as an *in vivo* model of a nerve-muscle preparation, and as a dynamically developing and changing system. Observations have been made to elucidate the role of the nerve both in maintenance and in maturation of skeletal muscle.

EFFECTS OF BOTULINUM TOXIN ON SKELETAL MUSCLE

Materials and methods

Type A crystalline botulinum toxin (generously supplied by Dr. E. Schantz, Fort Detrick, Maryland), stored at 3°C in acetate buffer, was freshly diluted in chick embryo Ringer solution (Rugh, 1961) immediately before use. Each 0·1 ml. volume of the solution contained 30 to 60 μg. of toxin, an amount which represents more than 10,000 lethal doses for hatched chickens. The diluted toxin was injected directly into the chorioallantoic circulation of the experimental embryos by the following technique, described in detail elsewhere (Drachman and Coulombre, 1962*a*). A rectangular window was removed from the shell and shell membrane overlying the embryo, to permit access for injection and observation. A specially constructed microcatheter was inserted within a chorioallantoic vein, under a binocular dissection microscope. Each injection of 0·1 ml. was made with a micrometer-driven syringe attached to the catheter.

Over 300 chick embryos were used in these experiments. Several different dosage schedules were used, but all embryos received the initial injection of botulinum toxin on either the seventh or the 12th day of incubation. Since the timing of the first injection proved to be more important in determining the severity of the muscle changes than the other dosage parameters, the experimental embryos have been divided into two groups on this basis. *Group A* includes the embryos first treated on the seventh day of incubation; *Group B* includes those first treated on the 12th day. In some cases, the embryos received only one or two injections, while others received as many as six doses. Control embryos were given repeated injections of 0·1 ml. of Ringer's solution. The embryos were incubated at 37·7° in a humidified forced-draft incubator. All embryos included in the results were alive and in good condition, although paralysed, at the end of the experiments. They were killed on the 19th day of incubation, except where otherwise specified.

Twenty-four Group A and B embryos were used for weight determinations. The yolk sac and extra-embryonic membranes were removed, and the body was then washed, blotted and weighed. One thigh and leg with all attached muscles were dissected from each embryo, and skinned, blotted and weighed. After all soft tissue was removed, the femur, tibia and fibula were weighed; the difference between limb weight and bone weight was taken as the weight of the soft tissue. The soft tissue included skeletal muscle, connective tissue and fat.

Fixation for histological processing was carried out in 10 per cent formol-saline. After decalcification in formic acid-citrate solution, the thighs and legs were embedded in paraffin and sectioned serially at a thickness of 6 μ in the sagittal and transverse planes.

Cholinesterase staining was done according to the thiolacetic acid (Mumenthaler and Engel, 1961) and α-naphthylacetate (Pearse, 1960) methods, on material fixed overnight in cold $CaCl_2$-formalin, and sectioned at 45 μ in an International® cryostat.

RESULTS

The 19-day embryos injected with botulinum toxin were smaller and lighter in weight than the controls, but otherwise had matured normally for their age. Externally, they showed severe ankylosis of multiple joints (arthrogryposis multiplex congenita), a condition that has been shown to result from skeletal immobilization during embryonic development (Drachman and Banker, 1961; Drachman and Coulombre, 1962b). Apart from the additional finding of a slightly shortened upper beak, the abnormalities were limited to the muscular system. On gross observations, all skeletal muscles of the body were strikingly shrunken and fatty (Fig. 1).

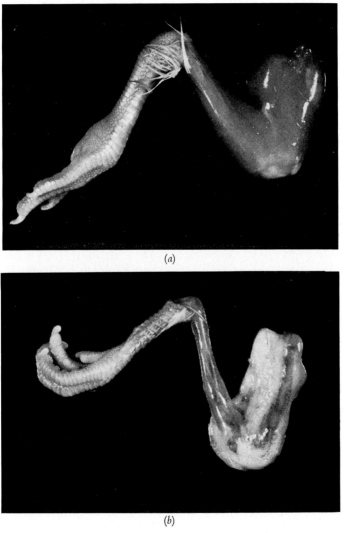

FIG. 1. BOTULINUM TOXIN. Limbs of 19-day-old chick embryos, with skin removed. (a) Normal control, with plentiful muscle. (b) Embryo treated with botulinum toxin from 7th to 19th days (Group A). Note severe loss of muscle, and replacement by fat (white-appearing material).

This appearance was more marked in the Group A embryos. The average limb muscle weight of ten embryos in this group was 0·204 g. (S.E. 0·007), a reduction of 85 per cent as compared with the normal (1·37 g., S.E. 0·053).

Histologically, the limb muscles showed atrophy, degeneration and fatty replacement, and these were more severe in the Group A cases. The most striking finding was the devastating reduction in muscle bulk (Fig. 2): most of the muscle was replaced by adipose tissue and often only a few strands

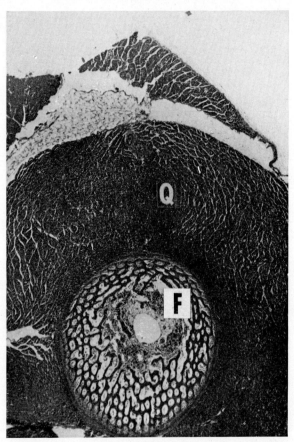

FIG. 2 (*a*)

FIG. 2. BOTULINUM TOXIN. Transverse sections through anterior thighs of chick embryos at 19 days. (*a*) Normal control chick embryo. (*b*) Embryo treated with botulinum toxin (Group A). At this low magnification, the dark area represents muscle; the light reticulated area is fat. Note the severe loss of muscle and massive replacement by fat in the botulinum-treated specimen. A small amount of atrophic muscle is indicated by the arrow. Haematoxylin and eosin. ×35 approx. F = femur; Q = quadriceps muscle.

of muscle could be identified in a sea of fat. The remaining muscle fibres were *atrophic, degenerating,* or *myotubal.* In these fibres there was an (apparent) increase in the number of sarcolemmal nuclei, with alignment, clumping and pyknosis of nuclei. Some of the muscle fibres retained their striations while undergoing atrophy; others showed marked degener-

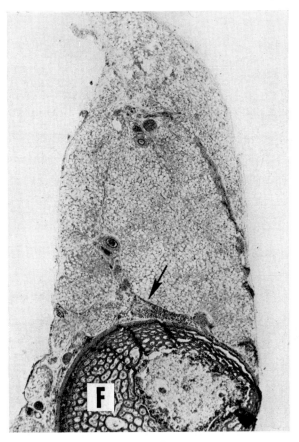

FIG. 2 (*b*)

ation, characterized by eosinophilia, swelling and floccular changes. Phagocytic histiocytes were present in the neighbourhood of degenerating fibres (Fig. 3). In many but not all cases, some normal-sized fibres were found scattered singly or in groups. In transverse sections, these fibres had the characteristic appearance of *myotubes,* with either a space or a nucleus in the centre of a ring of myofibrils (Fig. 4). Normally, myotubes are rare beyond 14 days of incubation age in the chick embryo. Their continued presence in the 19-day treated embryos indicates retardation of muscle maturation.

Spindle-shaped cells were found in some areas, scattered among atrophic muscle fibres. Examples of two or more spindle cells linked end-to-end were encountered. Their appearance suggested an origin from multi-nucleated skeletal muscle.

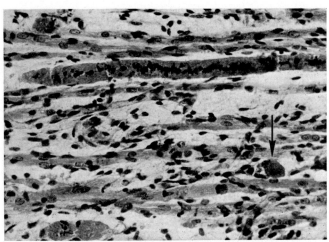

Fig. 3. BOTULINUM TOXIN. Thigh muscle of 19-day embryo treated with botulinum toxin (Group A). Note degenerating muscle fibre running transversely in upper portion of photo, histiocyte (arrow), and increase in sarcolemmal nuclei. H & E. × 320.

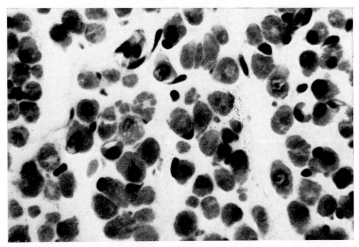

Fig. 4. MYOTUBAL MUSCLE. Transverse section of muscle from 19-day chick embryo treated with botulinum toxin (Group A). Note space or nucleus in centre of each fibre, surrounded by myofibrils. H & E. × 900.

The course of the muscle changes after various durations of intoxication was studied in 19 embryos killed at 11, 13, and 15 days of incubation age. The overall muscle bulk, as seen in transverse sections, was only slightly reduced at 11 days. The reduction was more marked at 13 days, and was striking at 15 days, by which time fatty replacement was also advanced. The proportion of individual muscle fibres showing atrophic and degenerative changes was small at 11 days, but very much greater in the 13- and 15-day specimens. The histological features were qualitatively the same as those described for the 19-day chick, but there was proportionately more muscle undergoing atrophy and degeneration and less fatty replacement at these ages. Particularly prominent in the 13- and 15-day specimens was degeneration of muscle fibres, with flocculent disintegration and eosinophilia of the cytoplasm, and phagocytosis by histiocytic cells. Spindle cells were more numerous at these ages. In the 15-day specimens, all the surviving muscle was myotubal.

Muscle spindles fared neither better nor worse than extrafusal muscle in the botulinum-poisoned material. In regions of massive fatty replacement, no muscle spindles were to be seen; in areas where some atrophic extrafusal muscle remained, the muscle spindles were comparably atrophic.

The histochemically-stained muscles showed absence or marked reduction in the number of cholinesterase-containing motor end-plates. Further investigation of the end-plates is currently in progress.

In summary, botulinum intoxication produced atrophy and degeneration of skeletal muscle in the chick embryo, which was noteworthy for the following features:
(1) The rapidity of the process, which was virtually complete within 12 days.
(2) The prominence of degenerative changes in muscle.
(3) The massive degree of fatty replacement.
(4) The retention of some primitive myotubal fibres.

EFFECTS OF CURARE ON SKELETAL MUSCLE

Like botulinum toxin, curare interferes with neuromuscular transmission. However, it acts via a different mechanism, which is described below (see Discussion).

In a series of experiments, intravenous infusions of *d*-tubocurarine (10 mg./ml.) were given to 24 chick embryos, beginning on the seventh day of incubation in 12 (Group A) and on the 12th day in the other 12 (Group B). The infusions were maintained continuously at a rate of 0·01 ml./hour until the 19th day of incubation, when the 15 surviving embryos were killed and fixed in 10 per cent formol-saline. The lower limbs were embedded in paraffin and sectioned transversely and sagittally.

The results were somewhat variable, but in the most severely affected cases the muscle atrophy and degeneration were closely similar to the changes seen after the other treatments. Fibre atrophy and degeneration, fatty replacement, nuclear changes, myotubes and spindle cells were all present (Fig. 5).

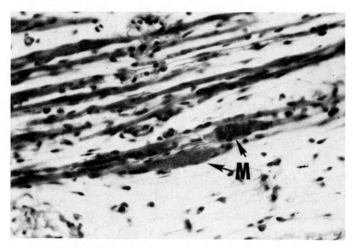

FIG. 5. CURARE. Calf muscle of 19-day embryo treated with curare (Group A). Note degenerating muscle fibre (M), spindle cells, and atrophic muscle fibres. H & E. × 400.

EFFECTS OF HEMICHOLINIUM ON SKELETAL MUSCLE

Hemicholinium is known to block neuromuscular transmission both pre- and post-synaptically. Its mechanisms of action, which are different from those of botulinum toxin and curare, are discussed later.

In this experiment, six chick embryos were given continuous intravenous infusions of HC-3 (generously supplied by Dr. F. W. Schueler) from the 12th through the 19th days of incubation. HC-3 was diluted in Ringer solution to a final concentration of 20 mg./ml., and infused at a constant rate of 0·01 ml. per hour. The five surviving embryos were killed on the 19th day of incubation, and fixed in 10 per cent formol-saline. The lower limbs were embedded in paraffin, and sectioned transversely and sagittally.

The severity of skeletal muscle atrophy and fatty replacement in the

HC-3 group was comparable to that after botulinum treatment (Fig. 6). Fibre atrophy, vacuolar and floccular degeneration, spindle cell formation and nuclear changes were also observed.

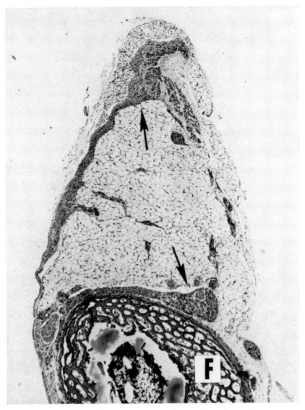

FIG. 6. HEMICHOLINIUM. Transverse section through anterior thigh of chick embryo treated with hemicholinium from 12th through 19th days. Note marked muscle atrophy and replacement by fat. F = femur. Arrow indicates muscle. H & E. × 35 approx.

CONTROL OBSERVATIONS

From the foregoing, it is evident that each of the pharmacological treatments produces dramatic histopathological effects on the skeletal muscle of the chick embryo. The following control observations were carried out in order to determine whether these effects were due to non-specific toxic or growth-retarding influences. Some of the control experiments apply only to botulinum toxin.

(1) *Control injections*: (a) *Ringer solution*: all embryos treated with repeated

injections of Ringer solution developed normally, and had histologically normal muscle. (b) *Toxin and antitoxin*: botulinum toxin was diluted 9 : 1 with specific antitoxin (Equine Antitoxin, Type A and B. Kindly supplied by Dr. Ruegsegger, Lederle Laboratories) instead of Ringer solution, and incubated at 37° for one to two hours before injection. Two doses of 0·1 ml. of the mixture were given to each of five embryos on the seventh and ninth days of incubation. None of the embryos was paralysed, indicating good neutralization of the toxin. When killed on the 19th day, all the embryos were grossly normal, with intact muscle and flexible joints. Limb muscle weights averaged 1·16 g., which is normal for that age. Histological sections showed normal muscle. This finding indicated that features of the infusion procedure other than the action of the toxin itself were not responsible for the muscle changes.

(2) *Effect of botulinum toxin on organ systems other than skeletal muscle*: (a) *Liver*: the weight of the liver was only 15·5 per cent below normal in the Group A embryos, as compared with a reduction of 85 per cent for the limb soft tissue. Since the growth rates of the liver and lower limb are normally parallel between the seventh and 19th days of incubation (Romanoff, 1960), it is clear that general retardation of all growth cannot account for the muscle findings. (b) *Heart*: although skeletal muscle was paralysed by the neuromuscular blocking agents, the cardiac striated muscle functioned normally throughout incubation. The total heart weight was only 25·7 per cent below normal in the Group A botulinum-treated embryos, and the cardiac muscle was histologically normal. This indicates that botulinum toxin does not exert a direct toxic effect on all *striated* muscle.

(3) *Absence of "general teratogenic" action of botulinum toxin, curare, and HC-3*: the pharmacologically-treated chick embryos showed no other deformities apart from the primary muscular and secondary skeletal abnormalities described. In contrast, gross defects in the central nervous system, head, eyes, rump, limbs, internal organs, feathers, etc., have resulted from treatment of *early* chick embryos with a wide variety of other drugs and cellular poisons (Zwilling, 1952; Ridgway and Karnofsky, 1952; Karnofsky, 1965).

SURGICAL DENERVATION IN THE CHICK EMBRYO

In order to compare the effects of the pharmacological treatments with those of denervation of the chick embryo, a simple procedure for extirpation of the entire lumbosacral spinal cord was devised.

Because of certain anatomical considerations, the operation was found to be most feasible at six days of incubation. A rectangular window was cut in the shell overlying the embryo, and the membranes were incised so as to permit access for surgical manipulation. A dorsal midline incision was made at the lumbosacral region, and the lumbosacral spinal cord was teased out with sharpened jewellers' forceps. After the operation, the lower limbs remained paralysed. The embryos were allowed to develop further *in ovo*, and were killed at 11, 13, 15 and 19 days of incubation age. Forty-

FIG. 7. SURGICAL DENERVATION. Lower limb of surgically denervated chick embryo, at 19 days. Note severe loss of muscle, and replacement by fat.

two lower limbs from 22 embryos were embedded in paraffin, and sectioned serially at 6 μ.

Grossly, the muscle was severely atrophic and replaced by fat (Fig. 7). This appearance was more marked in the older specimens.

Histologically, the skeletal muscle showed atrophic and degenerative changes, which were essentially the same as those seen in the pharmacologically-treated muscle. Fibre atrophy, nuclear changes, floccular degeneration and eosinophilia, phagocytosis, spindle-cell formation, muscle spindle atrophy, and the persistence of small numbers of atrophic or myotubal fibres, were all seen in the denervated material. The time course of development of these changes at 11, 13, 15 and 19 days corresponded closely to the course in botulinum-treated muscle, although it was slightly more advanced at each age. The present observations are in agreement with the findings reported by Eastlick in the muscle of non-innervated grafted limbs of the chick embryo (Eastlick, 1943; Eastlick and Wortham, 1947).

The virtual identity of the histopathology of denervated and pharmacologically treated muscle strongly supports the interpretation that the neuromuscular blocking agents faithfully reproduce the effects of denervation in the chick embryo.

THE INTEGRITY OF MOTOR NEURONS

It is important to determine whether botulinum toxin, curare and HC-3 exert their actions in the chick embryo by altering the *function* of the motor nerves, or by directly damaging them structurally. For this purpose, the spinal cords of botulinum-treated embryos were studied histologically. (Similar studies in embryos subjected to the other treatments are not yet available.)

Transverse sections were cut semi-serially, at a thickness of 6 μ. Counts were made of the number of intact motor nerve cells in the ventrolateral cell columns of segments 25–28. This region of the spinal cord is readily identified in birds by a prominent "glycogen body" which lies between the dorsal horns (Romanoff, 1960; Bueker, 1943). The large cells of the ventrolateral column are concerned with motor function in the lower limbs. This column of cells is not divided at segmental boundaries, but runs continuously through the vertical extent of the lumbosacral enlargement.

In the *Group B* embryos, the neurons were normal in appearance and number as compared with controls. This observation established that muscular atrophy can result from botulinum intoxication *while the motor neurons are structurally intact*.

However, in the Group A embryos there was a definite reduction of ventrolateral column cells. The remaining cells showed central chromatolysis, with cellular rounding, peripheral location of nuclei, and prominence of nucleoli. This finding is consistent with the phenomenon of retrograde degeneration of motor neurons in the embryo and young neonate after destruction of the peripheral field of innervation. Appropriate anterior horn cells degenerate within three to four days after limb bud amputation in the early chick embryo (Bueker, 1943; Hamburger, 1958), and within seven days after individual muscle destruction in the newborn mouse (Romanes, 1946). In the present material, the atrophy of muscle in Group A embryos was so profound as to lead in turn to secondary degeneration of the motor neurons. In the Group B embryos, the muscle atrophy, although severe by the end of the experiment, presumably had not yet had time to produce recognizable retrograde degeneration of the motor neurons. Further studies are under way to determine the time course of the anterior horn cell changes.

These observations are supported by the appearance of the axons in botulinum-treated embryos, as seen in Gros-Bielschowsky silver prepara-

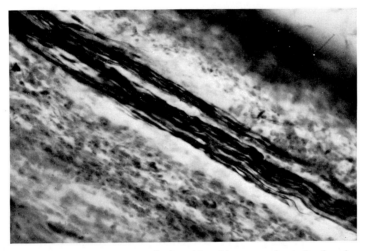

FIG. 8. INTRAMUSCULAR NERVE, in the midst of degenerated muscle of 19-day embryo treated with botulinum toxin. Axons are preserved, with slight swelling in some areas. Gros-Bielschowsky method. ×240.

tions. Even in the most severely affected specimens, intramuscular nerves were present, although probably in reduced density. Branches were seen in the vicinity of muscle which was undergoing degeneration (Fig. 8). In the areas of severest fatty replacement, there was beading of nerve fibres.

DISUSE OF MUSCLE

Like older animals, chick embryos are paralysed by the neuromuscular blocking agents used in these experiments. The question arises of whether lack of movement *per se* was the cause of the muscle atrophy and degeneration. In order to determine the effect of disuse on the chick embryo, an amputation-tenotomy operation was devised.

The procedures were carried out on the seventh to ninth days of incubation. After the air sac had been punctured, a rectangular window was cut in the shell overlying the posterior portion of the chick embryo. The right leg (which lies near the surface) was grasped with forceps and delivered through the opening in the membranes. The limb was severed at the knee or ankle joint with iridectomy scissors, no provision being made for haemostasis. The embryo was replaced within the amniotic sac, the window was sealed with cellophan tape, and the egg was returned to the incubator. The 44 surviving embryos were killed on the 19th day of incubation.

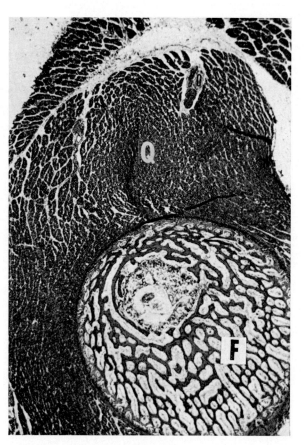

FIG. 9 (a)

FIG. 9. TENOTOMY. Transverse sections through anterior thighs of one operated chick embryo. (a) Left limb, unoperated side. (b) Right limb, after patellar tenotomy operation. Note mild degree of quadriceps atrophy. F = femur; Q = quadriceps muscle. (Print in b has been reversed, for comparison.) H & E. ×35 approx.

Since the quadriceps femoris muscle takes origin from the femur and inserts in the patellar tendon (Chamberlain, 1943), knee amputation is tantamount to quadriceps tenotomy. Similarly, amputation through the ankle severs the tendon of the peroneus longus muscle, which originates from the tibia (tibio-tarsus bone). Amputation-tenotomy relieves these muscles of their usual work load, thus permitting relative disuse. In

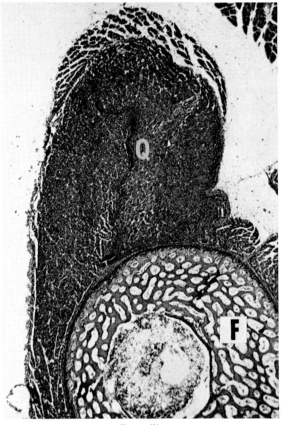

FIG. 9 (b)

embryos which were subjected to knee amputations, the patella was found to be drawn proximally up the thigh, indicating a shortened state of the quadriceps muscle. The shortened posture is optimum for producing the atrophic effects of tenotomy, according to Eccles (1943).

Sections cut transversely through the thigh or leg revealed only moderate decrease in the bulk of the quadriceps and peroneus longus muscles respectively, as compared with the unoperated side (Fig. 9). Neither marked fibre atrophy nor degeneration was found in the tenotomized muscles.

DISCUSSION

This study was designed to explore the hypothesis that acetylcholine may be essential for transmitting a trophic influence of motor nerves to developing skeletal muscle. We have found that prolonged treatment of chick embryos with three different pharmacological agents which block the cholinergic transmission system resulted in severe atrophy and degeneration of the muscle. This process is remarkable for its speed and completeness in the experimental embryos. Very little identifiable muscle remains, the rest having been replaced by masses of fat. Certain other histological features are particularly characteristic of the pharmacologically-treated muscle, including the retention of immature myotubal fibres, and the presence of spindle-shaped cells.

This picture reproduces in faithful detail the histological findings seen after surgical denervation of skeletal muscle in the chick embryo. Not only is the final appearance of denervated and pharmacologically-treated muscle virtually identical, but the time course of the atrophy and degeneration runs parallel after surgical denervation and botulinum intoxication. Other possible explanations of the histological changes include (*a*) general retardation of embryonic growth and development, (*b*) direct toxic damage to striated muscle, or (*c*) other non-specific effects of the experimental procedures, which have been considered and rejected, in a series of control experiments. This evidence provides a firm basis for the conclusion that botulinum toxin, curare and HC-3 actually produce denervation atrophy of skeletal muscle.

The effects of these agents are not due to primary destruction of motor neurons or axons. In the botulinum experiments, the motor nerve cells and their processes remain histologically intact at a time when muscle atrophy is advanced. This is consistent with the findings of Thesleff (1960), who reported that prolonged local treatment of cat and frog muscle with botulinum toxin did not damage the ultrastructure of the nerve endings. An interesting incidental observation in botulinum-treated chick embryos is the retrograde degeneration which *later* affects the motor neurons. This is most likely due to loss of the "peripheral field of innervation" without which the motor cells of embryonic or young animals degenerate (Bueker, 1943; Hamburger, 1958; Romanes, 1946).

It is also unlikely that disuse of muscle secondary to the prolonged neuromuscular blockade can account for the histological changes observed. In order to explore this possibility, a tenotomizing operation, which provides a fair approximation of disuse, was carried out as part of the present study. Chick embryos with patellar tenotomies showed only a mild degree of atrophy of the quadriceps muscle, not comparable to the effects of the pharmacological treatments. This finding is consistent with previous observations in adult animals, that the consequences of disuse of muscle are less complete and rapid than those of denervation (Tower, 1937).

Thus, it is concluded that botulinum toxin, curare and HC-3 are capable of producing denervation atrophy of embryonic skeletal muscle, by virtue of their pharmacological actions.

The interpretation of any pharmacological experiment depends on a close analysis of the agents used. The three agents employed in the present study are all highly purified crystalline substances, which share in common only the property of interfering with cholinergic transmission. They are chemically unrelated, and act by entirely different mechanisms, at different sites in the cholinergic transmission system. The principal sites of action of botulinum toxin, curare and HC-3 are illustrated diagrammatically in Fig. 10.

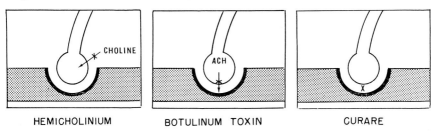

FIG. 10. CHIEF ACTIONS OF NEUROMUSCULAR BLOCKING AGENTS. Diagrams represent neuromuscular junctions. X indicates predominant site of blocking action of each agent. Hemicholinium blocks uptake of choline by nerve ending, and interferes with synthesis of ACh. Botulinum toxin blocks release of ACh at nerve terminal. Curare blocks access of ACh to end-plate receptor. For details, see text.

The Type A botulinum toxin used in the present study is a crystallizable protein which is produced by a strain of the anaerobic bacterium, *Clostridium botulinum* (Duff et al., 1957). It has been shown to be a specific inhibitor of both spontaneous and impulse-directed ACh release. Botulinum toxin in fractional microgram quantities causes lethal paralysis in most animals (Burgen, Dickens and Zatman, 1949; Brooks, 1956; Lamanna, 1959). Its only other known action, haemagglutination *in vitro* (Lamanna, 1948), appears not to be significant in the intact organism.

Curare (*d*-tubocurarine) is a pure alkaloid of known chemical composition and structure (Bryn Thomas, 1963). Its main pharmacological action is to block access of ACh to the receptor at the motor end-plate (Taylor, 1959). Curare has been shown to block the end-plate's response to ACh delivered via *spontaneous* neural release (miniature end-plate potentials), neural impulses (end-plate potentials) or experimental application (Fatt and Katz, 1952; del Castillo and Katz, 1957). Recently, it has been found to inhibit "post-tetanic repetitive activity" of motor nerves (Standaert, 1964).

The hemicholiniums are a class of hemiacetal compounds first synthesized by Long and Schueler (1954). Hemicholinium no. 3 (HC-3), the most intensively studied of these preparations, has been shown to prevent the synthesis of ACh at intact nerve endings by blocking the uptake of choline. This action may be antagonized by an excess of choline. In high concentrations, HC-3 also exerts a post-synaptic neuromuscular blocking effect (Hofmann, 1966).

It is most reasonable to conclude that impairment of ACh transmission, the common denominator of action of botulinum toxin, curare and HC-3, is responsible for the denervation atrophy they produce. By the same token, it would be implausible to assume that these strikingly different pharmacological agents all coincidentally possessed the property of blocking some other "trophic substance" in addition to their known effects on cholinergic transmission. It follows that ACh operates as a "trophic transmitter" in this system, and that there is no need to postulate the existence of any other substance to explain the motor nerve's ability to maintain the histological integrity of skeletal muscle.

The present studies go no further in specifying the mechanism of action of ACh as a "trophic transmitter." However, from previously reported work, it is improbable that ACh serves merely as an essential nutrient for muscle, which is normally provided at the motor nerve ending but might otherwise be supplied by some diffuse external source. Thus, Miledi (1960a) found that prolonged immersion of denervated frog muscle in a solution containing ACh failed to prevent the spread of the receptor zone.

Several published reports suggest that release of ACh by neural *impulses* is not essential for the trophic maintenance of muscle. Tower (1937) silenced the activity of motor nerves by surgically isolating the lumbosacral spinal cord segments of dogs. The muscle innervated by these quiet segments showed disuse atrophy which differed histologically from true denervation atrophy in several, but not all, respects. Denny-Brown and Brenner (1944) reversibly interrupted nerve conduction by compressing peripheral nerves, and noted the absence of muscle fibrillation. Similarly, Gutmann and Žák (1961) blocked conduction by maintaining a local anaesthetic infusion around the sciatic nerves of rabbits. Infusions of three days' duration failed to produce changes in muscle RNA, DNA and protein which would have occurred after denervation of equal duration. Other experiments also support the view that nerves which do not conduct propagated impulses may be capable of exerting trophic influences on muscle (Luco and Eyzaguirre, 1955; Emmelin and Malm, 1965). The most plausible possibility remains that the *spontaneous* quantal release of ACh by the nerve terminals normally conveys a trophic influence of motor nerves to skeletal muscle.

The most frequently cited evidence against the ACh hypothesis is based on a study of denervated frog muscle into which the motor nerves were allowed to regenerate. Miledi (1960b) called attention to certain disparities between the retraction of the chemosensitive zone and the presence of miniature end-plate potentials: (1) Several end-plates which had attained a normal frequency of miniature end-plate potential discharge still showed the enlarged zone of chemosensitivity characteristic of absence of the trophic effect of the nerve. (2) Conversely, in a small number of muscle fibres, the chemosensitive zone had retracted at a time when the frequency of miniature end-plate potentials was low. These observations have been interpreted to mean that spontaneous ACh release does not mediate the trophic effect of the nerve. On the contrary, the first situation is predictable on the basis of the ACh hypothesis, since the release of ACh by nerves would be expected to *precede* its effect on muscle. The second finding may be questioned because of the small number of observations. However, it would be of significance if it is substantiated by additional data.

Finally, it must be pointed out that the present analysis pertains only to the trophic system of motor nerve and skeletal muscle, and not necessarily to the many other situations in which nerves exert trophic effects on innervated organs.

SUMMARY AND CONCLUSIONS

(1) Repeated injection or infusion of three neuromuscular blocking agents, botulinum toxin, curare and hemicholinium-3, were made directly into the chorioallantoic circulation of chick embryos.

(2) After treatment of seven to 12 days' duration, the skeletal muscle showed atrophic and degenerative changes which corresponded closely to changes produced by surgical denervation of chick embryos.

(3) Control experiments were carried out which indicated that the histological findings were not attributable to:

(a) incidental effects of the infusion or injection procedure;
(b) "general retardation" of growth;
(c) primary destruction of motor nerves, axons or muscle, or
(d) disuse of muscle.

(4) It is concluded that botulinum toxin, curare and hemicholinium-3 produce pharmacological denervation atrophy of skeletal muscle in the chick embryo. Arguments have been presented which favour the view that acetylcholine transmits a trophic influence from motor nerves to muscles.

Acknowledgements

Mrs. Stephanie McLean and Miss Karen Luebbers provided technical assistance. A preliminary report of part of this work appeared in *Science* (Drachman, 1964; copyright by the American Association for the Advancement of Science), and in *Archives of Neurology* (Drachman, 1967).

REFERENCES

BROOKS, V. B. (1956). *J. Physiol., Lond.*, **134**, 264–277.
BRYN THOMAS, K. (1963). *Curare, its History and Usage*, p. 144. Philadelphia: Lippincott.
BUEKER, E. O. (1943). *J. exp. Zool.*, **93**, 99–129.
BURGEN, A. S. U., DICKENS, F., and ZATMAN, L. J. (1949). *J. Physiol., Lond.*, **51**, 10–24.
CHAMBERLAIN, R. W. (1943). *Atlas of Avian Anatomy*. Lansing: Hallenbeck.
DEL CASTILLO, J., and KATZ, B. (1957). *Proc. R. Soc. B*, **146**, 357–361.
DENNY-BROWN, D., and BRENNER, C. (1944). *Archs Neurol. Psychiat., Chicago*, **51**, 1–26.
DRACHMAN, D. B. (1964). *Science*, **145**, 719–721.
DRACHMAN, D. B. (1967). *Archs Neurol., Chicago*, **17**, 206–218.
DRACHMAN, D. B., and BANKER, B. Q. (1961). *Archs Neurol., Chicago*, **5**, 77–93.
DRACHMAN, D. B., and COULOMBRE, A. J. (1962a). *Science*, **138**, 144–145.
DRACHMAN, D. B., and COULOMBRE, A. J. (1962b). *Lancet*, **2**, 523–526.
DUFF, J. T., WRIGHT, G. G., KLERER, J., MOORE, D. E., and BIBLER, R. H. (1957). *J. Bact.*, **73**, 42–47.
EASTLICK, H. L. (1943). *J. exp. Zool.*, **93**, 27–49.
EASTLICK, H. L., and WORTHAM, R. A. (1947). *J. Morph.*, **80**, 369–389.
ECCLES, J. C. (1943). *J. Physiol., Lond.*, **103**, 253–266.
EMMELIN, N., and MALM, L. (1965). *Q. Jl exp. Physiol.*, **50**, 142–145.
FATT, P., and KATZ, B. (1952). *J. Physiol., Lond.*, **117**, 109–128.
GUTMANN, E. (1964). In *Mechanisms of Neural Regeneration*, pp. 72–106, ed. Singer, M., and Schadé, J. P. [*Progress in Brain Research*, vol. 13.] New York: Elsevier.
GUTMANN, E., and GUTTMANN, L. (1942). *Lancet*, **242**, 169–170.
GUTMANN, E., and HNÍK, P. (1962). In *The Denervated Muscle*, pp. 13–51, ed. Gutmann, E. Prague: Czechoslovak Academy of Sciences.
GUTMANN, E., and ŽÁK, R. (1961). *Physiologia bohemoslov.*, **10**, 493–500.
HAMBURGER, V. (1958). *Am. J. Anat.*, **102**, 365–409.
HOFMANN, W. W. (1966). *Ann. N. Y. Acad. Sci.*, **135**, 276–286.
JOSEFSSON, J. O., and THESLEFF, S. (1961). *Acta physiol. scand.*, **51**, 163–168.
KARNOFSKY, D. A. (1965). In *Teratology Principles and Techniques*, pp. 185–213, ed. Wilson, G., and Warkany, J. Chicago: University of Chicago Press.
KNOLL, P., and HAUER, H. (1892). *Sber. Akad. Wiss. Wien* III, **101**, 315. [Cited by Tower, S. S. (1939). *Physiol. Rev.*, **19**, 1–48.]
LAMANNA, C. (1948). *Proc. Soc. exp. Biol. Med.*, **69**, 332–336.
LAMANNA, C. (1959). *Science*, **130**, 763–772.
LEVI-MONTALCINI, R., and VISINTINI, R. (1938a). *Boll. Soc. ital. Biol. sper.*, **13**, 976–978.
LEVI-MONTALCINI, R., and VISINTINI, R. (1938b). *Boll. Soc. ital. Biol. sper.*, **13**, 938–985.
LONG, J. P., and SCHUELER, F. W. (1954). *J. Am. pharm. Ass.*, **43**, 79–86.
LUCO, J. V., and EYZAGUIRRE, C. (1955). *J. Neurophysiol.*, **18**, 65–73.
MILEDI, R. (1960a). *J. Physiol., Lond.*, **151**, 1–23.
MILEDI, R. (1960b). *J. Physiol., Lond.*, **154**, 190–205.
MUMENTHALER, M., and ENGEL, W. K. (1961). *Acta anat.*, **47**, 274–299.
PEARSE, A. G. E. (1960). *Histochemistry*, pp. 886–887. Boston: Little, Brown.
RIDGWAY, L. P., and KARNOFSKY, D. A. (1952). *Ann. N. Y. Acad. Sci.*, **55**, 203–215.
ROMANES, G. J. (1946). *J. Anat.*, **80**, 117–130.
ROMANOFF, A. (1960). *The Avian Embryo*. New York: Macmillan.
RUGH, R. (1961). *Laboratory Manual of Vertebrate Embryology*, 5th edn. Minneapolis, Minn.: Burgess.
STANDAERT, F. G. (1964). *J. Pharmac. exp. Ther.*, **143**, 181–186.
TAYLOR, D. B. (1959). *Anesthesiology*, **20**, 439–452.
TELLO, J. F. (1917). *Trab. Lab. Invest. biol. Univ. Madr.*, **15**, 101–199.
THESLEFF, S. (1960). *J. Physiol., Lond.*, **151**, 598–607.

THESLEFF, S. (1961). *Ann. N. Y. Acad. Sci.*, **94**, 535–546.
TOWER, S. (1931*a*). *Bull. Johns Hopkins Hosp.*, **48**, 115–125.
TOWER, S. (1931*b*). *Brain*, **54**, 99–110.
TOWER, S. (1935). *Am. J. Anat.*, **56**, 1–43.
TOWER, S. (1937). *J. comp. Neurol.*, **67**, 241–267.
ZWILLING, E. (1952). *Ann. N. Y. Acad. Sci.*, **55**, 197–200.

DISCUSSION

Singer: It looks as if all neurons produce trophic substances. Possibly these substances are produced primarily for the neuron itself for the maintenance of its own axonal processes and the peripheral effects are secondary: the neuron may produce so much trophic substance and with so much enthusiasm that it spills over into the periphery. In many of these trophic systems, including the one of limb regeneration in the salamander on which I have worked, there is no evidence that acetylcholine is the effective agent. For example, regeneration which does not occur in the absence of the nerve occurs in the presence of an isolated sensory system. Pharmacological assay shows no acetylcholine in these circumstances. In the evolution of the neuron two functions were elaborated: one is that of conduction and the other trophic maintenance of the axon and secondarily the end-organs. In the first acetylcholine is involved as a peripheral mediator in the motor neuron. Is it not possible that the acetylcholine mechanism and the trophic one are two facets of a single biochemical background that evolved in the neuron, and that botulinum toxin, which you have used so successfully to suppress the trophic action of the nerve, actually hits the common biochemical mechanism? This could explain the relation which you suggest between acetylcholine and trophic activity.

Drachman: Your question is one which has repeatedly been raised in connexion with Thesleff's experiments and our own early experiments using *only* botulinum toxin. The objection was that botulinum toxin might block some other "trophic" substance in addition to acetylcholine. In order to examine this very point we have used additional pharmacological agents which affect cholinergic transmission at *different* stages of the process. It is highly unlikely that all three agents, which are themselves so different chemically, and which have such different modes of action, could interfere with some other substance in addition to acetylcholine. In the light of this multiple pharmacological approach, it is most likely that only the acetylcholine system need be affected to produce atrophic effects.

Eccles: Why do you use such high doses of botulinum toxin?

Drachman: I wanted to interfere with *all* release of acetylcholine, and the high dosage of botulinum toxin was used to prevent the release of even a nanogram of ACh.

Eccles: But isn't there evidence that botulinum toxin eliminates all action of the nerve terminals on muscle, and that you have a pharmacological denervation mimicking in every respect a true denervation, so that you have more than just a pharmacological blockade of ACh liberation?

Drachman: Thesleff has published electron micrographs made by Professor B. Katz, showing that the nerve terminals are intact after prolonged botulinum treatment (Thesleff, S. [1960]. *J. Physiol., Lond.*, **151**, 598–607).

Eccles: S. Fex, B. Sonesson, S. Thesleff and J. Zelená (1966. *J. Physiol., Lond.*, **184**, 872–882) showed that during botulinum "denervation" the muscle fibres accepted innervation from an implanted nerve just as with a denervated muscle.

Gutmann: L. W. Duchen and S. J. Strich (1966. *J. Physiol., Lond.*, **189**, 2P) have shown that in botulinum intoxication there is a tremendous sprouting of axons and induction of new spots of cholinesterase. Have you ever seen such a thing, Dr. Drachman? You said that there is some cholinesterase activity in denervated muscle. As you did the experiment at a stage when cholinesterase is not yet formed, it would mean that in spite of blockage of liberation of acetylcholine, you did not block the cholinesterase synthesis.

Drachman: S. Thesleff, J. Zelená and W. W. Hofmann (1964. *Proc. Soc. exp. Biol. Med.*, **116**, 19–20) showed that function can be restored more rapidly to botulinum-paralysed muscles by crushing the motor nerves. The nerve then regenerates, and the new sprouts innervate the muscle and restore motor function.

Duchen and Strich (1966, *loc. cit.*) found remarkable sprouting of the terminal and pre-terminal nerve fibres in the muscles of botulinum-poisoned mice, as you mentioned.

From these two studies it appears that there is nothing wrong with the original nerve endings in the poisoned muscles; it is just that new endings sprout.

Eccles: How is it known there is nothing wrong? Is this done with the very best electron microscopic techniques?

Drachman: The electron micrographs done by Professor Katz look very good. The study of Duchen and Strich was done with a silver impregnation method.

Eccles: Your very high dosage of botulinum toxin puts this into a completely new kind of category about what happens to the nerve terminals.

Drachman: Lower doses produce the same effects (Drachman, unpublished results).

Eccles: R. Miledi (1963. In *The Effects of Use and Disuse on Neuromuscular Functions*, ed. Gutmann, E., and Hník, P. Prague: Czechoslovak Academy of Sciences) also was of the opinion that botulinum could be destroying not only the emission of acetylcholine but possibly also of other substances such as "mysterine", though of course he did not use that word!

Drachman: Professor Gutmann's question about cholinesterase is a complicated one. I have studied cholinesterase histochemically in botulinum-treated chick muscle. I find that at a time when discrete cholinesterase-containing end-plates are normally present in the chick embryo (that is at about 15 or 16 days) the botulinum-treated muscle is fairly severely degenerated. Only a few cholinesterase-containing end-plates can be seen in botulinum-treated muscle at these ages.

B. C. Goodwin and I. W. Sizer (1965. *Devl Biol.*, **11**, 136–153) found no significant difference between the cholinesterase activity of denervated limbs of chick embryos and those of normal embryos. They concluded that cholinesterase formation, at least up to the 16th day of development, is independent of innervation.

Gutmann: The authors are relating cholinesterase activity to protein content of a section of the muscle.

Drachman: The denominator is one of the problems.

Székely: Let met say a few words in favour of "mysterine", Dr. Drachman. There are two kinds of acetylcholine in the muscle tissue: myogenic and neurogenic. I do not know what the difference is between these two, but they show the same reaction in histochemical specimens. In studies of the development of myoneural junctions in chicken embryos, M. Mumenthaler and W. K. Engel (1961. *Acta anat.*, **47**, 274–299) and K. Straznicky (1967. *Acta biol. hung.*, **18**, 437) have found that the myogenic acetylcholine can be shown in the myoblast cells from the seventh day of incubation. The neurogenic acetylcholine appears first five days later in the form of small spots on the sarcolemma. From that time on the myogenic acetylcholine gradually disappears, but if the muscle develops under aneural conditions, this acetylcholine remains until the 16th day, when the muscle deteriorates. Now, if it is just the acetylcholine the muscle needs for normal differentiation, it has plenty.

Straznicky replaced the brachial segments of the spinal cord by thoracic segments in chicken embryos. The wing muscles degenerated and disappeared the same way as in a nerveless limb graft, although motor nerve fibres of thoracic spinal cord origin invaded the wing. Under the microscope the degenerating muscle fibres were rich in myogenic acetylcholine, but the thoracic motor fibres could not establish myoneural junctions and no neurogenic acetylcholine was observed. All the same, the muscle completely deteriorated in the presence of acetylcholine and motor nerve fibres.

I think these observations shift the problem of maintenance and differentiation of the muscle tissue towards a functional aspect, as Dr. Hník has proposed before.

Walton: In the Prostigmine-treated embryos of which you showed some slides, Dr. Drachman, there seemed to be some nuclear changes which were very much more suggestive of regeneration than of degeneration. This should be followed up. For instance, does the sarcoplasm around these nuclei stain with azure B for RNA, before and after predigestion with ribonuclease?

Drachman: I suspect that there is indeed a good deal of regeneration taking place. However, the rapid growth of embryonic muscle normally gives rise to a histological appearance which resembles regeneration in the adult.

Gutmann: It is interesting that the neuron can be maintained in your experiments without the production of acetylcholine, i.e. that regeneration of the neuron continues in spite of the blocking of acetylcholine production. You have apparently not knocked out all the chemical systems which are important for maintenance of neuron and muscle. Thus in spite of your very fine experiments the problem remains unsolved.

Drachman: The multiple pharmacological approach makes the possibility of interfering with systems other than those concerned with acetylcholine most unlikely.

Eccles: Were you proposing that curare (*d*-tubocurarine) prevents the release of acetylcholine?

Drachman: I really have no original thoughts on the subject. Its well-established effect is, of course, post-synaptic. F. G. Standaert (1964. *J. Pharmac. exp. Ther.*, **143**, 181–186) has written that *d*-tubocurarine inhibits post-tetanic repetitive activity in motor nerves by some presynaptic action.

Eccles: The generally accepted evidence is that tubocurarine doesn't alter acetylcholine release except in very high doses. It does influence presynaptic terminals but I don't think it has ever been shown to block or prevent production of acetylcholine. Your evidence is indirect and it could be interpreted in other ways.

Drachman: For the purposes of the present hypothesis, the argument is stronger if curare has only a post-synaptic action. I have mentioned a presynaptic action only because it has been described in the literature.

Eccles: You could say that these procedures prevent the acetylcholine from acting on the membrane across the synaptic cleft. But then you are up against the trouble that an acetylcholine atmosphere doesn't seem to work that way in Miledi's experiments (1960. *J. Physiol., Lond.*, **151**, 1–23). Undoubtedly it gets to the neuromuscular junction. The best way of giving a muscle acetylcholine might be to put it in an acetylcholine bath, as was done by Miledi. This is perhaps overdoing the ACh application, but it doesn't in fact have a trophic influence.

Kerkut: Maybe one could pulse it in discontinuously on and off?

Drachman: It would be appropriate at this point to consider *how* acetylcholine may exert its presumed trophic influence. In our preliminary experiments (unpublished) it appears that subjecting the muscle to an excess of ACh (by inhibiting cholinesterase via neostigmine) produces just as harmful an effect as depriving the muscle of all neural ACh. This suggests that the muscle's exposure to ACh must be intermittent rather than sustained, and is fully consistent with R. Miledi's observation just mentioned by Sir John (1960, *loc. cit.*). It may be that the significant "trophic" effect of ACh is merely one of repeatedly altering the permeability of the muscle membrane both to ionic and to non-ionic substances.

Eccles: There might be a sustained effect, of course, but it may not be working to give impulses.

Kerkut: At high concentration the acetylcholine might inhibit, as with the bell-shaped curve for cholinesterase activity in the presence of high acetylcholine levels.

Eccles: This has all got to be looked at very critically indeed. I feel myself attracted by "mysterine", because we know other things cross the junction. Nerve after all can change the speed of muscular contraction. Acetylcholine is not responsible for changing the muscle contraction times and all kinds of other detailed machinery inside the muscle fibre. Both the slow and fast muscles have acetylcholine operating in their nerve terminals. There are more things in the neuromuscular mechanism than acetylcholine transmission.

Drachman: I certainly agree to that.

Whittaker: Couldn't the function of the acetylcholine be to release mysterine? There has been a lot of talk about presynaptic actions of transmitters. One of its functions might be to release a trophic substance.

Eccles: But how does it release a slowly contracting muscle substance when the nerve terminal goes to a fast muscle? The nerve cells themselves apparently can change the muscle, and they don't do it by acetylcholine.

Crain: Instead of considering acetylcholine to be a "trophic neuromuscular transmitter", might it not be more accurate to refer to its role in facilitating the transport of various trophic agents from the neuron to the muscle cell? Acetylcholine may be important in getting the trophic agent across the muscle cell membrane, but it may not be a trophic agent itself. It may simply be facilitating transport of much more specialized neuronal macromolecules at synaptic junctions. Analysis of pharmacological denervation atrophy of skeletal muscle *in situ* should also take into consideration the fact that isolated skeletal muscle fibres can be maintained for months, or years, in tissue culture, even though no specific acetylcholine system may be involved. Perhaps trophic molecules analogous to NGF are present in the culture media. In the chick embryo, *in situ*, might it not be possible to distinguish between the actions of a post-synaptic agent like curare, and presynaptic agents like botulinum and hemicholinium, by seeing whether the embarrassment produced by the latter agents can be neutralized by perfusing acetylcholine into the embryo to replace that which is presumably not being liberated at the treated presynaptic terminals? Damage produced by curarization, on the other hand, would not be neutralized by such perfusion of acetylcholine.

Drachman: I agree that it is very likely that acetylcholine facilitates the transport of essential substances across the membrane. I cannot exclude the possibility that one of these substances is a specific "trophic agent" for muscle. However, if you accept the ACh-permeability hypothesis there is no logical necessity to postulate such an agent.

The fact that skeletal muscle in tissue culture, as well as cardiac striated muscle *in vivo*, can survive without the benefit of any neural trophic influence has intrigued me. One respect in which these systems differ from skeletal muscle *in vivo* is that they are capable of initiating spontaneous repetitive electromechanical activity. It is possible that this idiomuscular activity repetitively alters membrane permeability, and thus maintains the integrity of the muscle cells. By contrast, skeletal muscle is unable to initiate and maintain repetitive membrane activation *in vivo*, and therefore may require the motor nerve's ACh to do so.

As to the experimental use of acetylcholine, the first step before trying it was to perfuse the vascular system of the embryo with an anticholinesterase agent. This is necessary to prevent the rapid hydrolysis of acetylcholine. I have already mentioned the preliminary experiments in which Prostigmine produced muscle degeneration.

Whittaker: Prostigmine is not as specific an anticholinesterase as eserine or one of the organic phosphorus compounds.

Hník: I find it rather surprising that tenotomy produced such small changes in the muscle. Is it not possible that the muscles did not retract sufficiently but made contact with the bone? They could then exert at least some effective tension. Dr. J. Zelená (1963. *Physiologia bohemoslov.*, **12**, 30–36) tenotomized muscles in rat foetuses and newborn rats and the changes she found were much more dramatic than in your experiments.

GENERAL DISCUSSION

Szentágothai: In the study of the forces guiding axons to their respective targets, both in the periphery and the centre, three fundamentally different stages or mechanisms have to be distinguished: (1) start of axon growth from the neuroblast in a determined direction, (2) conduction along the trajectory of the fibre, and (3) arrival at the site of termination and selection of the appropriate cell or part of the cell to establish synaptic contact.

(1) I have discussed the possible significance of orientation of the axis of the neuroblast in the "aiming" of the axon in a determined direction earlier in the discussion (see p. 96, this volume). We know nothing about what might determine the orientation of neuroblasts—it may be chemical gradients or many other factors. It is fair to admit that the direction of primary outgrowth of the axon is not an absolutely determining factor. There are several experiments reported in the literature with rotation of nervous primordia around various axes (for example: Piatt, J. [1949]. *J. comp. Neurol.*, **90**, 47–93), with the result that many fibre tracts start to grow in the wrong direction and may keep up this wrong direction for long distances or indefinitely. However, certain fibres at least sooner or later readjust their courses and may reach their correct targets.

(2) As discussed lucidly on several occasions by P. Weiss (1950. Summarized in *Genetic Neurology*, ed. Weiss, P. Chicago: University of Chicago Press), conduction along the trajectory of the pathway depends on many factors, such as: ultrastructural orientation of the medium, contact guidance on cell surfaces or surfaces of blastemata, specific (glial) guiding structures (e.g. cells of the optic stalk, surface glial zones in various parts of the medullary tube), etc. Ultrastructural orientation of the tissue matrix occurs automatically between two neighbouring groups growing nerve cells. The interstitial tissue or space situated between two such foci will be oriented in the direction connecting the two centres of the foci, with the effect that axons grown out from both foci will be preferentially directed from one focus to the other. This effect has long been exploited in tissue culture. If it is correct that axons tend to maintain a straight direction during growth as long as they are not caused to deviate by some obstacle or preferential micellar orientation of the environment, one would expect that axons bridging the distance between two cell groups or blastema(s), and situated not too far from each other, ought to cross, i.e. those arising from one side of the focus ought to reach the other side of the focus and *vice versa*. We have termed this the "camera lucida" principle of neural projection (Szentágothai, J., and Székely, G. [1956]. *Acta biol. hung.*, **6**, 215–229). This fundamental principle of organization of crossing connexions is very common in invertebrate nervous systems, particularly in those lacking bilateral symmetry.

The case is peculiar in the vertebrate CNS with its strict bilateral symmetry, or—for that matter—in all nervous systems with bilateral symmetry. Here the growth and differentiation of the two halves of the medullary tube necessarily cause transverse tensions in the two connecting laminae, the roof plate and the basal plate. Any axon tip growing forward in the neighbourhood of one of these plates—if caught in this field of transverse stresses and consequently transverse ultrastructural orientation of the tissue matrix—will be easily led to cross to the other side. This appears to be an important factor in bringing about fibre commissures and fibre tract crossings (Szentágothai and Székely, 1956, *loc. cit.*). I do not propose to explain everything on such a mechanical basis; it is only that these aspects of development should not be left out of consideration.

(3) The selection of the exact target with which to establish synaptic contact upon arrival of the fibre at the region of termination might be explained by mechanisms such as those mentioned by Sir John (p. 96, this volume). Acceptance or rejection between axon terminal and cell surfaces on the basis of matching or non-matching molecular composition or other cues is entirely possible. We know both from tissue culture and from developmental studies that the growth cones of the axons send out finger-like temporary sprouts, which are withdrawn if the surface reached is inappropriate, and strengthened or further elaborated and established if the contact is appropriate.

Stefanelli: Basic factors affecting the orientation of nervous fibres are clearly shown in tissue culture, where some features of the living body are absent. In tissue cultures polarity disappears and also asymmetry between a left and right half. Moreover there are no gradients—as the cephalo-caudal and dorsal-ventral in an organized living system.

Eccles: But how does the fibre know about laterality?

Stefanelli: I will give you some experimental results. If explants of cerebellum or tectum present a free surface then there is an orientation of the fibres in respect of this surface, in the absence of which the cells become arranged differently, for example in rosettes with the fibres showing a radial orientation. In the tectum there are specific cells with "looping" fibres. It is possible to find these cells in explants of tectum only if the explants have a polarity, otherwise the looping fibre does not emerge. Crossing of the fibres is another example: there is an intrinsic tendency of specific fibres to go to the "other" side. In explants of the medulla containing a Mauthner cell, which develop bilaterally, fibres cross to the contralateral half; something induces the fibres to cross, independently of any functional significance.

Eccles: What is the information flow system?

Stefanelli: Different nerve cells have a genetic quality which is fixed in the very early stages when histogenetic determination takes place. When a neuroblast is histotypically determined, it manifests its own quality even when explanted. We might try to find out what are the forces that guide fibres. We might also postulate that the cells have different qualities with which to respond to a stimulus.

We can try to find out what forces are acting on the unity of a structure. The whole organism is a "unity", which a tissue culture is not. The behaviour of fibres in culture is thus of a basic type. In culture, we can't study the normal morphological forces within symmetrical and polarized systems.

Hughes: One interpretation of the experiments where a barrier is placed in the way of outgrowing nerve fibres is that normal development rests on very labile equilibria. Hamburger put a barrier between the cord and outgoing fibres in the frog tadpole (Hamburger, V. [1928]. *Arch. EntwMech. Org.*, **114**, 272–363). Such experiments were repeated by Pierre Tschumi and myself. The fibres will get round by the front or the back if they can, but such a barrier greatly reduces the number getting into the limb, and sometimes none get in at all. Under these circumstances what are the fibres doing, since they are produced in something like normal numbers from spinal ganglia? If we look at them from underneath, when they are frustrated from their normal goal in *Xenopus* they form bizarre anastomoses and completely transcend normal bilaterality: they are all over the place, forming quite perverse anastomoses. Such experiments indicate to me how very delicate the equilibria must be which determine normal neurogenesis (Hughes, A. F., and Tschumi, P. A. [1958]. *J. Anat.*, **92**, 498–527).

Eccles: When W. Harkmark (1955. *J. comp. Neurol.*, **100**, 115–209; 1956. *J. exp. Zool.*, **131**, 333–371) excised the lip of the rhombencephalon in the embryo, axons of cells from the embryonic pontine nuclei grew up along their normal pathways looking for the missing target (the cerebellum). Not finding it and not being able to make the proper synaptic connexions, these pontine cells died.

Kollros: S. Ingvar many years ago wrote a short paper on electrical fields or potentials of directing fibres (1920. *Proc. Soc. exp. Biol. Med.*, **17**, 198–199). In 1924 Weiss wrote a paper which essentially refuted this (*J. exp. Zool.*, **68**, 293–448). In 1946 G. Marsh and H. W. Beams wrote a paper (*J. cell. comp. Physiol.*, **27**, 139–157) on ganglia in culture which indicated very clearly that nerve fibres could grow in relation to a particular field—that is, direction of fibre outgrowth was influenced. The fact that crossed fibres are seen suggests that we can't simply depend upon the ultrastructural organization because it is followed by only one set of fibres. The other set clearly doesn't follow it. I have no notion whether there is in any embryo anything in the way of electrical fields of the calibre required to guide outgrowing fibres even over micro-distances. Nonetheless, we know that in the experimental situation the nerve fibres have this capacity of following.

Eccles: As regards neurobiotaxis, attempts to modify the rate of growing nerve fibres by electrical fields have given no significant results.

Hamburger: One of the most serious shortcomings in neuroembryology is that we do not have an inkling of an answer to the question: "What are the signals to which the tip of the growing axon responds?". We have to make efforts to envisage an experimental design that might begin to answer these questions. I think it is indicated in W. Harkmark's scheme (1955. *J. comp. Neurol.*, **100**, 115–

209) that the factors which determine the pathway to the goal are independent of the factors which are responsible for the synaptic connexions and for maintenance. This has come out very clearly in experiments in which typical pathways were established without subsequent connexions with the terminal organs. There are degrees of specificity and this may give us a hint of where to look and what kind of experiments to do. Professor Stefanelli has just brought out the very important point that the strain specificity of neurons is determined very early, so that each group of neurons acquires its biochemical specifications in early stages of differentiation. Also the matrix on which the fibres grow in the embryo must have a very complex pattern of biochemical specifications. "Pathfinding" is obviously an interaction of two biochemically very highly specified systems—the matrix and the axon with a whole spectrum of specificities of interaction.

One example of extremely low specificity comes from experiments by Dr. Piatt (1956; 1957. *J. exp. Zool.*, **131**, 173–202; **134**, 103–126). For instance, the thoracic spinal cord in the urodele amphibian can supply a reasonably good innervation for forelimbs; the nerve pattern is never perfect, as he has shown, but it approaches normal distribution, and if the cord is transplanted early enough innervation even results in function.

An example of an extremely high degree of specificity comes from experiments I did on the trigeminal nerve (1961. *J. exp. Zool.*, **148**, 91–123). The trigeminal ganglion has a double origin: from a placode and from neural crest. The nerves emerging from the placodal neurons, although they were in atypical positions, were perfectly capable of following every single major and minor pathway in the peripheral distribution pattern, whereas the neural crest compound of the trigeminal, which was much closer to these pathways, formed nerves which refused to enter these pathways.

The intermediate degrees of specificity are very well illustrated by the Mauthner cells, whose fibres apparently merely require information as to laterality and axial direction. They can follow pathways which are in a general way anterior/posterior but they do not require specific information as to whether they should grow near the ventral fibre tracts, which is their normal pathway. In experimental situations they may grow more laterally or even dorsally. They "know" simply that they have to grow caudally.

This spectrum of specific interactions has to be considered all the time. The degree of specification may be different for different sectors of the pathway: there may be high specificity at the end where synaptic connexion is established, but low specificity in the intermediate parts of the pathways. For example, sensory and motor fibres travel at first along the same highways and they then separate to go to different muscles and sense organs; obviously, they now follow specific signals which must be different from the more general signals which guided them while they travelled together.

Mugnaini: The cerebellum, because of our highly developed knowledge of its intrinsic organization, fibre connexions, and function of its units, is undoubtedly one of the most valuable models for the study of growth of nervous tissue in the central nervous system.

The most evident phenomena involved in cerebellar histogenesis are schematically the following: production of a certain number of undifferentiated cells; migration of undifferentiated cells; formation of furrows; differentiation of certain numbers of neuroblasts and glioblasts; degeneration of certain numbers of cells; migration of neuroblasts and glioblasts; direction of axonal growth (ipsilaterally, contralaterally, rostrally, caudally) and selection of the target areas; selection of target cells; modelling of pre- and post-synaptic patterns; attainment of a certain cell size and shape. Several of these phenomena have been dealt with by previous speakers, but the neuronal-glial and glial-neuronal relationships and the development of the vascular system, which also show different characteristics in the various regions of the central nervous system, have not been touched at all. In the cerebellum the relationships between nerve and glia cells are striking. Certain neurons (Purkinje cells and Golgi cells, the former being inhibitory neurons with short projecting plexiform axons) are "gliophile", while others are not (superior and inferior stellate cells, inhibitory neurons with short axons). Immediately around the Purkinje cells there is a very strong concentration of glia cells and during development most glia cells migrating up to the cortex stop and make their last mitotic division at the level of the Purkinje cells. All the receptive surface of the Purkinje cell not covered by synapses is covered by glia.

It seems that in general glioblasts differentiate later than neuroblasts.

In the cerebellar cortex there are very peculiar kinds of nerve cell populations and nerve fibre patterns, almost identical all over the cortex and highly similar in the various animal species. A detailed analysis of how its characteristic architecture is attained may therefore shed some light on several of the basic mechanisms in the growth of nervous tissue.

The *granule cells* are special intrinsic relay cells. They arise from cells in the so-called external granular layer. The axon is formed early. Most of the differentiation occurs in the absence of synaptic connexions. All the granules have a characteristic dendritic pattern and all the different afferent fibres contacting the granules form characteristic mossy endings. Collaterals of the mossy fibres which end in other regions do not form mossy endings.

Eccles: Do you mean that collaterals from these fibres branch out elsewhere?

Mugnaini: Yes. Some of the mossy fibres have collaterals ending outside the cerebellar cortex.

These facts suggest either that the granule neuron is a genetically well-determined cell which induces the synaptic pattern, or that in the granular layer there is a topographic (quantitative or qualitative) factor which determines synaptic modelling over mossy fibres or the dendrites of granule cells. (For further details see Mugnaini, E., and Forströnen, P. F. [1967]. *Z. Zellforsch. mikrosk. Anat.*, **77**, 115–143).

Basket cells apparently originate from the external granular layer. The neuroblast is oriented transversely to the folium and the axon is formed early. The cell nucleus lies near the axonal pole. The dendrites are formed after the axon has begun to grow and are early oriented in a single plane, transverse to the folium (perpendicular to the parallel fibres). (For the formation of baskets around the Purkinje perikarya, see Mugnaini, E. [1966]. *Anat. Rec.*, **154**, 391).

All the *Purkinje cells* in the mature cerebellum have a single dendrite which branches profusely in one plane, perpendicular to the parallel fibres. The latter synapse mostly with the peripheral portions of the Purkinje dendrite. The dendritic tree of the Purkinje cells is furthermore extensively contacted by the so-called climbing fibres, one of the most striking of all nerve fibres. These fibres may have short collaterals ending on stellate and Golgi cells. The fibres climb on the Purkinje dendrite, but not on the other cells (at least not typically).

The aim of my present work is to establish whether the climbing fibre plays some role in determining the arrangement of the Purkinje dendrite. The orientation of the Purkinje tree in a single plane is attained in the chick embryo at about 16 to 17 days of incubation. At this stage there are some characteristic climbing fibres at the level of the Purkinje dendrite. The problem can also be attacked experimentally with lesions in the centres of origin of the climbing fibres before differentiation.

In cerebellar cultures or in abnormal cerebella containing few granule cells the Purkinje cells are unordered. The parallel fibres, which are the first well-ordered element during development of the cerebellum, could well also be an important element in the organization of the Purkinje dendrite.

Many morphogenetic processes may be determined by genetic properties of the cells, as mentioned by Professor Stefanelli, as well as by contact guidance and subtle unknown topographic factors. Transplantation experiments have shown how these may interact. The establishment of cell-to-cell relationships and the modelling of the synaptic pattern most probably are governed mainly by specific chemical and electrical properties of the cells and their membranes and should be the object of biochemical and biophysical investigations.

Prestige: A hypothesis about how specific connexions could be established stems from work on the size of cells (Henneman, E., Somjen, G., and Carpenter, D. O. [1965]. *J. Neurophysiol.*, **28**, 560–580; McPhedran, A. M., Wuerker, R. B., and Henneman, E. [1965]. *J. Neurophysiol.*, **28**, 71–84). These workers have shown that in any pair of spinal motor neurons, the larger neuron will have a higher reflex threshold and a larger periphery than the smaller neuron. As a result large neurons have a lower usage than small neurons. Let us therefore speculate that the difference in size between large and small neurons is caused by use itself, that is, increased use of a cell causes it to become larger (hypertrophy), so that its threshold goes up and its use is restrained at the new cell size. This is a negative feedback hypothesis. Conversely, with decreased use, let the cell get smaller (atrophy), so that its threshold goes down and its use is facilitated. Let us follow through this hypothesis on a system of cells connected initially at random. When a cell fires it activates other cells. If any of the output cells is in negative feedback onto the first cell, the first cell is going to be used less than other cells with similar synaptic bombardments, because its own activity inhibits itself. Its usage is therefore less than the others and by the hypothesis its size gets smaller and its threshold goes down, until the decreased threshold balances the effect of the negative feedback. It is then smaller and it therefore has to have a smaller peripheral field. Which of its synaptic output contacts is it going to lose? If it loses negative feedback contact, then the loop is broken and the whole cycle starts once again. So that is not a stable solution. If it loses one of the others, the position is stable. Thus in a cell with an output that feeds back negatively onto itself, the size and threshold will alter and the peripheral field will be restricted selectively to retain those parts that cause the negative feedback. It will selectively lose the others. Conversely, if the cell has a positive feedback on itself, its usage is greater, and by hypothesis it gets larger, and its threshold is raised until it balances the excess use caused by the positive feedback. Since it grows larger, it also increases its peripheral field. If it picks up a negative feedback output, it will selectively retain it, as before, and it will drop positive feedback or neutral contacts. On the other hand, if it picks up another positive feedback output, it continues the process, and grows until it picks up a negative feedback contact, onto which it will again restrict, losing neutral or positive feedback contacts.

Thus on the basis of the hypothesis that there is a causal relationship between use and size, cells starting with random outputs will alter until their outputs are in negative feedback on themselves. The nervous system is conspicuous for the use of negative feedback in its construction. There are various ways to use this mechanism: it could be used in ontogeny, perhaps in connexion with spontaneous activity; it could be used to describe a way in which local or recurrent inhibition allows a whole pattern of nerve cells to be projected onto another set and the pattern retained. One could also make models of this type which show conditioning.

Gutmann: A simpler relation between fast and slow nerve cells and fast and slow muscle fibres can be found in their protein metabolism. The slow muscle fibres apparently have a higher turnover of proteins (Gutmann, E., and Syrový, I. [1967]. *Physiologia bohemoslov.*, **16**, 232) than fast muscle fibres and they are connected with small neurons which also have probably a higher turnover of proteins than the neurons innervating fast muscles. Thus we have a higher turnover of proteins of slow units, and this metabolic characteristic will be important whenever specificity and feedback systems are considered.

★ ★ ★

Eccles: We shall now turn to some of the clinical problems of growth of the nervous system.

Walton: We all appreciate that in learning in the central nervous system certain very complicated processes are involved, and obviously some very complicated neuronal pathways. It has been well recognized for many years that in the course of development certain specific learning defects may occur. One of these is developmental dyslexia, which has been shown in certain families to be genetically determined. The exact anatomical or physiological substrate of this kind of learning defect is completely unknown, although some evidence suggests that it tends to arise in individuals in whom dominance is not completely or fully established in one or the other cerebral hemisphere. In other words it arises in individuals in whom there is either left-handedness or in a broader sense some confusion in cerebral dominance. Many other forms of specific learning defect have been known to occur from time to time but it has only recently been recognized that the acquisition of motor skills may be similarly disturbed in children who demonstrate no evidence whatever of neurological abnormality in the broadest sense, and no other sign of brain damage. These are not spastic children; they are children who otherwise appear clinically to be completely normal but who have the greatest difficulty in acquiring motor skills. We have seen a number of these children and have tested them in detail over the last few years. After we had described the initial series of five (Walton, J. N., Ellis, E., and Court, S. D. [1962]. *Brain*, **85**, 603–612), we were astonished to find how common this defect is. A recent survey of schoolchildren in Cambridge, carried out by workers in Dr. Zangwill's department, has demonstrated that up to 10 per cent of otherwise normal schoolchildren have some degree of difficulty in acquiring movement and skilled motor activity (Brenner, M. W., Gillman, S., and Farrell, M. F. [1967]. *J. neurol. Sci.*, in press). The children we have tested are ones in whom the degree of motor clumsiness and the difficulty in acquiring movement was so great that it had seriously interfered with their development, to the extent that they had been regarded by skilled schoolteachers as being mentally backward. These children have come to light particularly in professional families, not because the condition is commonest in professional families, but because these children were subjected to educational pressures greater than those imposed upon children of artisan or working-class families.

These children are usually late in walking and in acquiring all types of motor skill. At school, although they may learn to read quickly and easily acquire verbal and often numerical ability, they are extremely slow in learning how to write and draw. Their clumsiness is so great as to constitute a severe handicap. The most important single diagnostic feature in these children is that most of them show a gross discrepancy between the verbal and performance I.Q. as shown on the Wechsler intelligence scale for children. The first case, a boy of 10, had been diagnosed by his schoolteachers as mentally backward, but he had a performance I.Q. on the Wechsler test of 97 and a verbal I.Q. of 137. However he found it completely impossible to dress or tie a necktie, for example. In other respects he was neurologically completely normal. In the group of 24 children we have now examined there are other defects (Gubbay, S. S., Ellis, E., Walton, J. N., and Court, S. D. [1965]. *Brain*, **88**, 295–312). Some children not only demonstrated clumsiness but a kind of fidgetiness which we have called "searching movements". Sometimes mirror movements are seen on the contralateral side. It is important to recognize that not only do the children have difficulty in acquiring motor skill or praxis but many have similar defects on the cognitive as distinct from the executive side. They demonstrate difficulty in learning how to appreciate and to interpret sensory information.

Buller: Is the mirror writing connected with dyslexia?

Walton: Some of them do show mirror writing, persisting for very much longer than that occasionally seen briefly in apparently normal children. One boy's movements were so gross that he had in fact been diagnosed initially as suffering from chorea. Many of these children show the kind of disorganized movement that Prechtl has described (Prechtl, H., and Beintema, D. [1964]. *The Neurological Examination of the Full-term Newborn Infant.* London: Heinemann). When one boy attempted to draw a bicycle he joined the front wheel to the rear by a chain, he then drew a spoke and put in the front handlebars not too badly; he then drew in the saddle suspended in mid-air from the chain midway between the front and the rear wheels.

Another boy aged 9 years, 9 months, had a verbal I.Q. of 97 but a performance I.Q. of only 44. When he was first sent to school he was so poor at finding his way around that another child had to take him from one class to another. When he copied writing patterns he had little or no sense of shape or direction. An even more severe defect was that he had great difficulty in recognizing the parts of the body on himself and on other children. At first he couldn't identify his own ear, for example, and he had the greatest difficulty in carrying out any kind of constructional activity. He was not aware of his own failures.

A third child had a lower I.Q. still and showed visual agnosia. Like the last boy, she had no concept of shape.

These children were incapable of benefiting from normal education; they were regarded as mentally backward and had to be specially trained, yet they did improve with increasing maturity; it is very difficult to find an adult who is equally clumsy.

I think this kind of abnormality is far more common than is generally realized. We are not sure whether it is due to pathological change within the parietal lobe of the brain concerned with the acquisition of skills, or to a physiological defect in organization. A high proportion of these children have confusion in cerebral dominance—they may be right-eyed, left-handed and perhaps right-footed. Minor degrees of developmental apraxia and agnosia are extremely common and may cause serious educational difficulties, but many of these children improve greatly with the passage of time, although those more severely disabled remain greatly handicapped. One can train many of these children by sheer repetition—going over and over again the movements which they have not succeeded in acquiring spontaneously.

Eccles: The important point arising from what you have said is that we are very apt to take a properly grown nervous system for granted. We have to realize the immensity of the problems as we try to imagine them from our present still primitive viewpoint. Before we knew anything about the problems these were all solved for us in some magical way by "mysterine" or what you will. In a very primitive way, we have been trying to understand this immensely complicated organization of basic information. This sets us a fantastic problem. Where do these changes that you have been able to show in these defective states occur? I should like to think that the cerebellum is involved. We have still to learn how such a wonderful performance results from a properly organized nervous system, and what a great struggle people not endowed in this way have to become active and useful members of society.

There we must leave our discussions for the present, with the hope that they will prove of value to all of us, and also to those who later read the book.

INDEX OF AUTHORS

Numbers in bold type indicate a contribution in the form of a paper; numbers in plain type refer to contributions to the discussions.

Angeletti, P. U. **126**

Benitez, Helena H. . . . **148**
Bornstein, M. B. **13**
Buller, A. J. 73, 107, 229, 230, 247, 248, 287

Crain, S. M. **13,** 36, 37, 38, 39, 145, 175, 177, 195, 196, 231, 277

Drachman, D. B. 75, 95, 107, 119, 123, 125, 144, 145, 146, 218, 232, 246, 248, **251,** 273, 274, 275, 276, 277

Eayrs, J. T. 143, 146, 192, 196
Eccles, Sir John **1,** 37, 39, 68, 70, 71, 72, 73, 74, 76, 95, 96, 97, 98, 107, 108, 119, 120, 121, 122, 123, 124, 145, 146, 147, 218, 230, 231, 246, 248, 249, 250, 273, 274, 276, 277, 280, 281, 283, 286, 288

Gaze, R. M. 38, **53,** 69, 71, 72, 75, 76, 122, 177, 217, 230
Gutmann, E. 94, 118, 124, 143, 232, **233,** 245, 246, 247, 248, 249, 274, 275, 285

Hamburger, V. 69, 98, **99,** 105, 106, 107, 108, 109, 121, 281
Hibbard, E. **41**
Hník, P. 36, 94, 108, 125, 176, 231, 243, 246, 247, 278
Hughes, A. 69, 70, 73, 76, 93, 95, 98, 105, **110,** 117, 118, 119, 120, 122, 123, 197, 216, 232, 250, 281

Kerkut, G. A. 36, 93, 107, 119, 142, 176, 198, **220,** 229, 230, 231, 232, 250, 276

Kollros, J. J. 69, 73, 95, 107, 117, 119, 121, 122, 124, **179,** 194, 195, 196, 197, 198, 199, 281

Levi-Montalcini, Rita 76, 108, 122, **126,** 143, 144, 145, 146, 147, 174, 195, 247

Mugnaini, E. 108, 219, 232, 282, 283
Muntz, Louise 105, 106, 176
Murray, Margaret R. 38, 146, **148,** 174, 175, 176, 177, 216, 247, 250

Peterson, E. R. **13**
Piatt, J. 94, 95, 117, 195
Prestige, M. C. 75, 118, 120, 121, 125, 198, 199, 249, 250, 284

Singer, M. 72, 123, 124, 147, **200,** 215, 217, 218, 219, 273
Sperry, R. W. 39, 40, **41,** 67, 68, 69, 70, 71, 72, 73, 74, 75, 76, 97, 98
Stefanelli, A. 31, 39, 196, 197, 215, 280
Székely, G. 37, 39, 68, 70, 71, 72, 73, 76, **77,** 94, 95, 97, 98, 108, 119, 121, 122, 197, 198, 275
Szentágothai, J. **3,** 67, 68, 72, 73, 74, 75, 96, 106, 143, 144, 146, 175, 194, 217, 229, 245, 246, 248, 249, 279

Walton, J. N. 73, 106, 122, 218, 245, 275, 286, 287
Watson, W. E. **53**
Whittaker, V. P. 277, 278

Young, J. Z. 69, 217, 218, 229, 230, 232, 247

INDEX OF SUBJECTS

Acetylcholine, action of, 276–277
  effect of botulinum toxin, 2
  effect of curare, 259–260, 276
  effect of hemicholinium, 260–261, 269
  as trophic neuromuscular transmitter, 251–278
    blocking of, 252, 269, 276
    in denervated muscle, 237
    release, of, 270
    types in muscle tissue, 275
Actin filaments, 118
Amino acids,
  penetration into axon, 202–205, 210
  penetration into myelin sheath, 201–202, 210–211
  penetration into Schwann cytoplasm, 201–202, 206
Androgens, effect on muscle, 241
Antuitrin G, effects of, 188
Axon,
  amino acid penetration, 202–205, 217
  degeneration of, 211–212
  dependence on cell body, 200, 207, 211–212, 213–214, 218–219
  development of, 112, 113, 114, 115, 216
    differentiation by size, 115
  growth of, 4, 5, 49–50, 279
    direction, 96–97
  metabolic input, 213–214, 219
  metabolic pathways to, 207–209, 227, 231, 247
  plasmic flow from cell body, 205, 211–212, 215, 218–219
  protein synthesis in, 210, 217
  relation to cell, 212–213, 227

Behaviour, 6
  and feedback, 68
  effect of hormones, 189, 193
Blastema, 124,
  growth of, 72
Botulinum toxin,
  effect on liver, 262
  effect on neurons, 264–265, 268

Botulinum toxin,
  effect on skeletal muscle, 253–259, 268, 273–275
  site of action, 269
Brachial segments of cord, interchange with thoracic segments, 79–80, 95, 111, 275
Brain,
  cell division and migration after optic nerve lesions, 53–67
  effect of growth hormone on development, 188
  neuron connexions with medulla or spinal cord, 19–28
  of embryos,
    effect of thyroid hormones, 187–188
  transport to muscle, 220–225
    effect of cold or Xylocaine, 222–223, 227, 229, 230, 231
Brain cells, in chick embryo, 101
Brain stem, neuronal connexions with spinal cord, 15–19

Cell division,
  in chick embryo, 107
  in motor neurons, 117–118, 120, 121
  in retina, 61–64
  in tectum, 55–61, 69
  in ventral horn, 112, 113–114
Central nervous system, transport to muscle, 220–225
Cerebellum,
  basket cells, 284
  effect of deuterium, 163, 176
  histogenesis in, 283
Cerebral cortex, effect of growth hormone, 195–196
Cerebrum, effect of deuterium, 166, 169–170
Chick embryo,
  degeneration in, 121
  effect of neuromuscular blocking agents, 252 et seq.
  movements in, 99–109
    during hatching, 101–102, 103, 104, 106

SUBJECT INDEX

Chick embryo,
    "goal-directed", 106
    periodic spontaneous, 99–101, 105, 108
    pre-hatching, 102–103, 107
    sensory control, 103, 105–106, 107, 108
    muscle twitching, 177
    "pipping", 103, 104
    retinal cells in, 34–35
Children, learning problems in, 286–288
Chromatolysis, and protein synthesis, 245–246, 247, 249 250
Clinical problems, 286–287
Cold, effect on transport from CNS to muscle, 222–223, 227, 229, 230
Colour perception, and optic nerve degeneration, 43
Contact guidance, 38, 97
Cornea, regeneration of, 72, 73
Corneal reflex, effect of thyroid hormones, 180–181, 197
Cranial nerves, specificity, 70
Curare,
    effects on skeletal muscle, 259–260, 276
    site of action, 269
Cutaneous fibres,
    labelling of, 69
    specific regrowth, 73

Degeneration,
    and protein synthesis, 121–122
    effect of testosterone, 119
    effect of thyroid hormones, 198
    mechanism of, 249–250
    of axons, 211–212
    of end-plate, 247
    of nerve fibres, 215–216
    of neurons, 120–125
    peripheral, 120–121, 123
    reasons, 120, 121, 122
Dendritic spines, and memory, 8
Deoxyribonucleic acid,
    in denervated muscle, 236, 241, 246
    synthesis in retina, 61
Deuterium,
    action of, 172–173, 175, 176
    on behaviour, 148
    on cerebellum, 163, 176
    on cerebrum, 166, 169–170
    on dorsal root ganglia, 161–163
    on growth of nervous tissue, 148–178

Deuterium,
    on hypothalamus, 163, 172
    on microtubules, 171–172, 175–176
    on mitosis, 171–172, 174
    on myelination, 163, 170
    on neuron, 155, 156–160, 163
    on salivary glands, 160–161
    on sympathetic ganglia, 150–160, 172, 175
Diencephalon, ependymal cells in, 60
Double dependence of growth process, 7

Embryo,
    effect of nerve growth factor, 127, 144
    limb grafting in, 115, 119
    movements in,
        in chick, 99–105
        in other species, 105–106
        pharmacology, 107–108
Endocrine influences in neural development, 179–199
End-plates,
    degeneration of, 247
    double, 248
    potential, effect of nerve section, 248
Epidermal growth factor, 141, 146

Function,
    and arrangement of motor neurons, 83
    and neural connexions, 42
    changes in, correlation with structure, 8
    of neurons, correlation with number, 91–92, 94, 95

Ganglia,
    dorsal root, effect of deuterium, 161–163
    effect of nerve growth factor, 132–136, 143, 147
    sympathetic, effect of deuterium, 150–160, 172, 175
Gastrocnemius, regeneration, 116
Glutamate, transport along nerve, 220–226, 230, 231–232
Gradient theory, 43, 74–75
Growth hormone,
    effect on cerebral cortex, 195–196
    effect on developing brain, 188

Hemicholinium,
    effect on skeletal muscle, 260–261
    site of action, 269

SUBJECT INDEX

Hormonal influences in neural development, 179–199
Hormones, effect on muscle, 240–241
Hypothalamus,
  effect of deuterium, 163, 172
  effect of thyroid hormone, 194

Inhibition, and CNS explants, 39–40

Learning,
  and growth hormone, 196
  and neurons, 41
  in nervous system, 286–288
  role of feedback, 68
Limb,
  innervation,
    development of, 110–125
    pharmacological aspects, 114
  movements of, 113
    and number of neurons, 94, 95
    development of, 77–95
    nervous control, 78–93, 95
  muscles,
    contraction sequence, 83
    effect of replacing brachial by thoracic segments, 79–80, 95, 111, 275
    represented in brachial segments, 81–85
    studied in movement, 84–85
  regeneration, 124
  specificity of nerve fibres, 69, 70
Limb grafting, 72
  homologous response, 111, 119
  innervation, 111, 115, 117
  movements, 119
  nervous control of, 78–79
Liver, effect of botulinum toxin, 262

Median eminence, effect of thyroid hormone, 189
Medulla, neuron connexions with brain, 19–28
Memory and neuron growth, 8
Mesencephalic fifth nucleus, action of thyroid hormones, 184
Microtubules, effect of deuterium, 171–172, 175–176
Mitosis,
  effect of deuterium, 171–172, 174
  effect of thyroid hormone, 187–188, 199

Motor nerve fibres,
  and limb movements, model experiments, 85–92
  innervating individual muscles, 81, 90–91
  diameters of regenerated, 94
  selection of peripheral pathways by, 79–81
Motor neurons,
  degeneration, 122–125
    causing decline of activity, 101
  in chick embryo, 107
  numbers of, 117–118, 120, 121, correlation with limb movement, 91, 92, 94, 95
  of ventral horn, 112, 113–114, 120, 122
  primary and secondary, 95
Muscle,
  activity, 77
    in chick embryo, 99–109
  atrophy,
    caused by lack of movement, 265–267
    due to botulinum toxin, 256–258, 268, 273–275
    due to curare, 259–260
    due to hemicholinium, 261
    due to surgical denervation, 262–264
    effect on sensory outflow, 243–244, 246
    integrity of neurons, 264–265
  cell membrane, transport across, 277
  denervated, 251
    by surgical means, 262–264
    protein synthesis in, 236–237, 245, 246
  effect of hormones, 240–241
  homologous response, 111, 119
  innervation of, 90–91
  limb,
    development of, 112, 118
  maintained by proteins from nerve cells, 236, 238, 240
  neurotrophic relations with nerve, 233–250
  skeletal,
    effect of botulinum toxin, 253–259, 268, 273–275
    effects of curare, 259–276
    effect of hemicholinium, 260–261
  species differences in metabolism, 245
  spindles, effect of botulinum toxin, 259
  spontaneous activity, 108

SUBJECT INDEX

Muscle,
  striation of, 118
  synergist-antagonist relationship, 81, 84
  transport to, from CNS, 220–225
    effect of cold or Xylocaine, 222–223, 227, 229, 230, 231
  transport to nerve, 225–226, 229
  twitching, 177
  with double end-plates, 248
Muscle-nerve contact, 119
Muscle receptors, effect of atrophy, 244, 246–247
Myelin,
  effect of deuterium, 163, 170
  effect of thyroid hormones, 195
Myelin sheath,
  amino acid penetration, 201–202, 210, 217
  relation to neuron, 212–213
  RNA in, 219
  structure, 206, 208, 213, 216
  transport in, 205–210, 211–212, 213, 215, 217
Myosin, 118
Mysterine, 274, 275, 277

Nerve,
  effect of strychnine on development, 16, 25, 39, 114
  neurotrophic relations with muscle, 233–250
  transport from muscle, 225–226, 229
  transport of material along, 220–232
    effect of cold or Xylocaine, 222–223, 227, 229, 230, 231
Nerve cells. *See under* Neurons
Nerve connexions,
  blocking of, 30
  formation of, 13–14, 98, 177–178
    electrophysiology, 15, 16, 17, 21, 23, 25–28
  in dissociated cells, 33, 38
  mechanism, 96, 280, 284–285
Nerve fibres
  See also under types, Optic nerve fibres, Motor nerve fibres, etc.
  chemical stimulation, and growth, 2
  counting, 122
  degeneration of, 215–216
  direction of growth, 96
  effect of cold on transport, 222–223, 227, 229, 230

Nerve fibres,
  effect of nerve growth factor, 136–138, 141, 143, 144
  effect of thyroid hormones, 115, 119
  growth of, 279–281
    contact guidance, 38, 97
    mechanism, 96
  orientation of, 279–280
  pathways, 110
  regeneration, 74–76
  size of, 115–116
  specification, 69–70
  without cell body, 215
Nerve growth factor, 126–147, 149
  action of, 127, 128
  discovery of, 126
  effect at subcellular level, 138–140
  effect on ganglia, 132–136, 143, 147
  effect on nerve fibre outgrowth, 136–138, 141, 143, 144
  nature of, 128, 129, 146
  production in mouse submaxillary gland, 126, 128–132, 140
  purification, 131–132, 145
  response to, "halo effect", 132–136, 145, 175
  specificity of, 145
  types of, 146
Nerve-muscle contacts, 119
Nerve pathways,
  formation of,
    possible mechanism, 5, 96
    role of specific cell surfaces, 98
Nerve processes, growth of, 4
Nerve section, protein loss in, 233–236, 245, 246
Neurobiotaxis, 41
Neuroblasts, orientation of, 279
Neuromuscular transmission, blocking of, 253–261, 269–270
Neurons,
  artificial, 87
  connexion and function, 42
  correlation of number and function, 91–92, 94, 95
  degeneration, 120–125
  differentiation, 4
  effect of deuterium, 155, 156, 157–160, 163
  effect of nerve growth factor, 127
    at subcellular level, 138–140
  effect of thiourea, 185

SUBJECT INDEX

Neurons,
  effect of thyroidectomy, 183
  effect of thyroid hormones, 181-183
  functional deprivation, 7-8
  genetic properties, 280, 284
  granule, 283-284
  growth of memory and, 8
  guiding mechanisms, 5
  in learning processes, 42
  in muscular atrophy, 264-265, 268
  Mauthner, 50, 282
    effect of thyroid hormones, 185-186, 196
  mitosis, 174
  modulation theory, 77
  plasmic flow to axon, 205, 211-212, 215, 218-219
  protein synthesis in, 233-236
  regeneration, 176
  relation to axon, 227
  relation to myelin sheath, 212-213
  saturation factors, 6
  size of, as factor in function, 284-285
  specificity of, 282
  synaptogenic properties, 28, 29, 37, 38
  trophic activity of, 212-213, 219, 273

Optic fibres,
  affinity with tectal neurons, 43
  correlation between tectum and retina, 45, 46
  deflection of, 47-49
  distribution in tectum, 49
  growth of, 46-50
  overlap on tectum, 44
  pathways,
    following retinal regeneration, 65, 76
    to tectum, 46-48
  regeneration, 44, 74-76, 110
    to tectum, 53, 69, 73-74
  terminal arborizations, 44
Optic nerve lesions, 42
  cell division in retina after, 62-64, 65
  cells in tectal ependyma, 55-61, 69, 75
  cells in tectal white matter, 55, 69
  cerebral cell division and migration after, 53-67
  re-learning after, 67-68
  tectal cell division in, 66

Perikaryon, 28, 29, 37, 121
  trophic role, 233

Peripheral nerves, development, 114-115, 124
Peripheral nerve fibres, 115, 120
  amino acid penetration, 200-219
Phototropism, 176
Phyone (somatotropic hormone), 188
Pigment epithelium, importance of, 35-36
Prostigmine, 278
Protein, transport in nerve, 227
Protein synthesis,
  and chromatolysis, 245-246, 247, 249, 250
  and degeneration, 121-122
  in denervated muscle, 236-237, 245
  in nerve cells, 233-236, 245-246
Purkinje cells of cerebellum, 284

Regeneration, 273,
  after crushing, 176
  cell recruitment during, 118
  of limbs, 124
  of retina, 62-64, 65
Re-learning, after optic nerve section, 67-68
Resonance principle of Weiss, 42, 97, 98
Retina,
  cell degeneration in, 122
  cell division in, 61-64, 65
  ciliary margin, 63, 64, 65
  growth of, 61-62
  importance of vitamin A, 35-36
  reaggregation of cells, 34-35, 38
  regeneration, 62-64, 65
  stimulation, 1
  stratification compared with tectum, 44-46
Retino-tectal connexions, 5
  after optic nerve section, 65, 66, 73-74, 75
  orderly factors in growth, 41-52
  regeneration of, 71
Ribonucleic acid in myelin, 219

Salivary glands, effect of deuterium, 160-161
Schmidt-Lantermann cleft, function of, 207, 208, 210, 216, 217-218
Schwann cytoplasm,
  amino acid penetration, 201-202, 206, 210
  constituents, 217-218
  pathways to axon, 207-209

Sensory outflow from muscles, effect of atrophy, 243–244, 246
Specificity, 5
    of nerve fibres, 69–70
Spinal cord,
    brachial segments,
        limb muscles represented in, 81–85
        replaced by thoracic segments, 79–80, 95, 111, 275
    neuronal connexions with brain stem, 15–28
    Rohon-Beard cells, 197
    role in control of limb movement, 78–79, 85, 93, 95
    structure, 6
Submaxillary gland of mouse, production of nerve growth factor in, 126, 128–132, 140
Strychnine, effect on developing nerve 16, 25, 39, 114
Synapses,
    and memory, 8
    differentiation, 36
    formation of, 5, 28–30, 37, 38

Tectum,
    cell division in, 55–61, 66, 69
    elasticity of gradients, 43
    ependyma, cells in, 55–61, 69, 75
    grey matter, cells in, 60
    optic nerve fibre distribution, 43, 49
    overlap of optic fibres, 44
    regeneration of optic fibres to, 53, 69, 73–74, 76
    stratification, compared with retina, 44–46
    white matter, cells in, 55, 69

Testosterone, effect on degeneration, 119
Thiourea, effect on neuron, 185
Thyroidectomy, effect on nervous system, 183, 193
Thyroid hormones,
    action on mesencephalic fifth nucleus, 184
    effect on behaviour, 193
    effect on corneal reflex, 180–181, 197
    effect on degeneration, 198
    effect on embryonic brain, 187–188
    effect on hypothalamus, 194
    effect on Mauthner cell, 185–186, 196
    effect on median eminence, 189
    effect on mitosis, 187–188, 199
    effect on myelin, 195
    effect on nerve fibres, 115, 119
    effect on nervous tissue, 189–190, 192
    effect on neural development, 179–180
    effect on Rohon-Beard cells, 197
    effect on ventral horn, 181–183, 195, 198
Tissue culture,
    of central nervous system, 13 et seq.
    synaptic formation in, 29, 37–38

Ventral horn,
    development of, 111–112, 113, 120, 122, 123
    effect of thyroid hormone on development, 181–183, 195, 198
Vitamin A, and retina, 35–36

Xylocaine, effect on transport from CNS to muscle, 222–223, 227, 231